SHIPHANDLING FUNDAMENTALS

— FOR —

LITTORAL COMBAT SHIPS AND THE NEW FRIGATES

TITLES IN THE SERIES

THE U.S. NAVAL INSTITUTE
BLUE & GOLD PROFESSIONAL LIBRARY

For more than 100 years, U.S. Navy professionals have counted on specialized books published by the Naval Institute Press to prepare them for their responsibilities as they advance in their careers and to serve as ready references and refreshers when needed. From the days of coal-fired battleships to the era of unmanned aerial vehicles and laser weaponry, such perennials as *The Bluejacket's Manual* and the *Watch Officer's Guide* have guided generations of Sailors through the complex challenges of naval service. As these books are updated and new ones are added to the list, they will carry the distinctive mark of the Blue and Gold Professional Library series to remind and reassure their users that they have been prepared by naval professionals and meet the exacting standards that Sailors have long expected from the U.S. Naval Institute.

Photo 1 *Freedom*-variant Littoral Combat Ship *USS Freedom* (LCS 1) pictured
U.S. Navy photo by Mass Communication Specialist 3rd Class Johans Chavarro

Photo 2 *Independence*-variant Littoral Combat Ship *USS Coronado* (LCS 4) pictured
Photo by author

SHIPHANDLING FUNDAMENTALS

FOR

LITTORAL COMBAT SHIPS AND THE NEW FRIGATES

JOSEPH A. GAGLIANO

NAVAL INSTITUTE PRESS
ANNAPOLIS, MARYLAND

Naval Institute Press
291 Wood Road
Annapolis, MD 21402

Library of Congress Cataloging-in-Publication Data
Gagliano, Joseph A., 1974-
  Shiphandling fundamentals for littoral combat ships and the new frigates / Joseph A. Gagliano.
      pages cm. — (U.S. Naval Institute blue & gold professional library)
   ISBN 978-1-61251-822-0 (alk. paper) — ISBN 978-1-61251-923-4 (ebook)
1. Littoral combat ships—United States. 2. Frigates—United States. 3. United States. Navy—Officers' handbooks. 4. Ship handling—Hand-books, manuals, etc. I. Title.
  V883.G34 2015
  623.88'25—dc23

                                                                2015023442

23 22 21 20 19 18 17 16 15   9 8 7 6 5 4 3 2 1
First printing

FOR SAM . . . *father and patriot*

Contents

TABLES

PHOTOS

FIGURES

In 1955, Capt. Russell Crenshaw published the first edition of what would become the fleet standard for shiphandling references, *Naval Shiphandling*, in which he foresaw usefulness in the experimental "hydrojets" of his day but was left only to imagine the possibilities. He conjectured, "Though hydrojets are only available today in sizes suitable for boats and small, special-purpose ships, it is probable that we will see much more of them in the future."[1] His predictions have come true with the U.S. Navy's construction of the littoral combat ship.

Waterjet propulsion is new for small surface combatants in the U.S. Navy, but certain shiphandling principles remain regardless of the type of propulsion moving the ship. This book is not intended to upend the authoritative references on naval shiphandling. In addition to Crenshaw's *Naval Shiphandling*, Capt. James Barber's *Naval Shiphandler's Guide* has served as an important update for this important field of study. It discusses such modern shiphandling training tools as simulators and when published provided a reference that was more stylistically approachable to younger generations of shiphandlers. While these two seminal works constitute the basic foundation for most shiphandling tasks, they are based primarily on propeller-rudder configurations and accordingly do not consider the principles and practical application of waterjets in shiphandling. This book is a fusion of concepts both old and new. To be sure, the shiphandling fundamentals of Crenshaw and Barber endure, so the chapters that follow are also not intended to reinvent them for the sake of innovation. Instead, this book builds on their lasting value by applying them in a practical way to help shiphandlers gain mastery in driving waterjet-equipped ships.

One theme that emerges throughout this book is the need to balance between the art and science of shiphandling. Chapter One begins this discussion by arguing that fanatical allegiance to either camp overlooks the ultimate goal of shiphandling—to maneuver the ship safely using the best tools for the task. Sometimes, careful mathematical preparation is required before a shiphandling evolution to understand better the shiphandling plan, but during the task itself shiphandlers do not often find themselves calculating equations. Some in fact might see this book as a deliberate movement away from the scientific approach, but a closer reading will reveal that the methods presented in the following chapters aim to strike a smarter approach that blends the two, using the science of shiphandling to build a stronger foundation for understanding but also for exercising the art of shiphandling when actually driving the ship. When the ship is in motion, its conning officer should focus on handling it, not calculating its theoretical movement.

The methods presented in subsequent chapters were developed over many years—during coursework in Newport, Rhode Island, experimentation on board ship, and in the simulator—but nothing was more valuable in their development than the training of novice shiphandlers. Watching them struggle to learn this skill and then giving them certain rules of thumb and methods for dealing with the challenge produced the methods and techniques that follow. In other words, the methodology of this book centers on junior officers with very little experience, because if they can use these methods to handle the ship safely, then even the most experienced shiphandlers should find them useful.

It is important to acknowledge, however, that the most senior shiphandlers reading this book, commanding officers, will have a much different perspective. The commanding officer's relationship with shiphandling is intensely personally, owing particularly to two factors. First, having somewhere around twenty years of commissioned experience, the commanding officer likely has certain methods or approaches that have served well in the past. The captain would not have achieved command at sea without a solid record of shiphandling, so it is to be expected that the methods underlying that record would act as the foundation for command. Second, safe shiphandling is one of the unambiguous performance markers that starkly determines whether a command tour is successful or not; it is only human nature for a self-preserving commanding officer to be skeptical of unfamiliar concepts.

While the methods presented in this book are tested and proven, they are not the *only* methods that could be employed. These methods are recommended because of their clarity and simplicity, but the captain's prerogative to drive the ship as he or she sees fit is an inviolable principle of command at sea. Captains seeking proven shiphandling techniques for this ship class can rely on these methods for both routine and emergency situations. For those preferring to forge their own methods, perhaps this book will provide a starting point for devising them.

For readers of this book who are junior officers reporting to their first littoral combat ships or new frigates—you should first be commended for reading about shiphandling *before* arriving on board. This book will prepare you to use waterjet propulsion to make the ship do truly extraordinary maneuvers. Still, you should keep in mind that understanding your commanding officer's preferred shiphandling method is of paramount importance for the ship's safe handling. When arriving on board or following a change of command during your tour, take time to study the captain's approach to shiphandling. Clear expectations between the shiphandler and the captain are critical in managing the inherent risk of shiphandling.

Finally, it should be noted that this book arrives during the initial stage of this ship class, as its design and designation are evolving. The Navy recently decided to incorporate improvements to the combat systems in a subsequent flight of this class and with those modifications will change its designation from "littoral combat ship" to "frigate." Ships with these modifications in their original build will be commissioned as frigates, and littoral combat ships receiving these combat systems upgrades will be given a frigate designation. For the shiphandler, however, these alterations will not considerably affect the ship's propulsion; the methods presented here will apply to both littoral combat ships and the new frigates.

One can only imagine the number of professionals involved in the development of the methods and techniques presented in this book, and most of these people likely had no idea that their everyday diligent efforts would someday contribute to it.

First, the collaboration that existed between the early commanding officers of littoral combat ships made them what remains the most productive peer group that I have encountered in twenty years of commissioned service. Whether meeting formally for staff meetings or informally for our monthly poker nights, we were constantly trading professional notes to compare methods that we developed individually on our own ships. As a commanding officer of an *Independence*-variant ship, I was particularly grateful for conversations with counterparts on *Freedom*-variant ships, from whom I learned that the similarities between these waterjet-driven variants were enough to warrant a single reference for both. Concurrently, these exchanges also identified those areas where the *Freedom* variant differed enough to demand procedural differences. Among this group of commanding officers, I am particularly thankful for conversations with Commanders Shawn Johnston, Dale Heinken, Rich Jarrett, Hank Kim, Matt Kawas, John Kochendorfer, Dave Back, and Jeremy Gray. I am especially indebted to Warren Cupps, who allowed me to observe his shiphandling team both in the simulator and under way, so that I could understand how the *Freedom* variant handles during pierwork. Moreover, I could not have worked through many of the mental exercises necessary to codify the maneuvers and principles in this book without the help of Mike Smith, my executive officer and successor on board *Independence*. I am indebted to him for serving as a sounding board in countless, and what may have seemed endless, conversations about shiphandling.

Second, the development of these methods began with an introductory course at the Surface Warfare Officers School in Newport. Comprising a group of civilian high-speed-shiphandling specialists—Dave Kane, Bill Lyons, Mike Howard, and John Parker—this team provided the foundation on which the subsequent chapters were developed. In addition to this course, the periodic training that occurred during off-hull periods in the simulator in San Diego, California, were invaluable for practicing certain methods before taking them to sea on what was then a one-of-a-kind ship. I especially valued the lengthy conversations with Robert Butt, a retired Navy captain and former cruiser commanding officer, who possesses immeasurable experience at sea and could provide extraordinary insights gained by observing every crew as it filed through the littoral combat ship (LCS) simulator. He is particularly well versed in the mathematics behind shiphandling, and I could always rely on him to talk through a new maneuver, method, or technique. As cited in this book, Bob deserves full credit for bringing forward certain concepts of driving waterjet ships. For example, what he refers to as the "balanced approach" directly led to the development of the *toe-in/toe-out method* presented in Chapter Three on waterjet vector management.

Third, I am deeply indebted to the officers and crew of USS *Independence*, without whom this book would not be possible. I could not confidently describe the methods and techniques presented in the following chapters as "simple and repeatable" without having taught them to successive groups of junior officers and senior enlisted sailors. Methods proven only by me would have had limited value, since they would only prove effective for shiphandlers with backgrounds and experiences similar to mine. For example, several tools developed early on, although logically sound and helpful to me because of my individual shiphandling experience, did not connect universally with those who needed them most. I was able, by observing these shiphandlers' successes and failures, to gauge the effectiveness of these tools, to adjust them through trial and error with the first shiphandling team of junior officers and senior enlisted sailors, and then to prove them over time with follow-on teams. Ships are often referred to as "living laboratories," but the shiphandlers on *Independence*—Bill Golden, Sam Bryant, Scott Tollefson, Roger Gonzalez, Wayne Lileks, Brent Hodge, Jordan Kelly, Lauren Howard, Will Henry, Kyle Decker, Nate Jones, and Tim Taton—served collectively as the real proving ground for this book.

This project was several years in the making, and although the research began long before a single sentence was composed, I credit my motivation to two specific individuals. First, while I was still the executive officer on board *Independence*, I worked for a commanding officer who gave me great latitude in driving the ship and teaching shiphandling to division officers. Matt Jerbi was very flexible in allowing me to experiment with *his* ship, and I still admire his accommodating nature. After one docking evolution in Port Hueneme, California, one that happened to go particularly smoothly, Matt leaned toward me and said, "You really should write an article about shiphandling." This small comment was very impactful, because it gave me confidence that I had something to contribute to the discourse on safe shiphandling. Second, during one of our many late nights in the simulator in San Diego, I happened to overhear Senior Chief Tim Taton comment to a fellow shiphandler, "We are watching the book get written on LCS shiphandling." I was pleased by the confidence that the sailors were gaining in these methods and techniques, but I am not sure that I would have taken on the challenge of this book project without this unintended nudge. Taken together, Matt Jerbi and Tim Taton helped me understand the need for this book. Matt gave me the confidence that I had something to contribute, and Tim identified the gap in the literature.

Finally, on a personal level, three particular individuals come to mind. Adm. Mike Mahon, one of my first commanding officers in the Navy, has been the most influential officer in my professional development. While I was still very young, he gave me the freedom to learn shiphandling "hands on," a gift that has paid dividends throughout the years. I am also thankful for Dr. Randy Papadopoulos—a good friend, confidant, and perpetual encourager—who

provided a critical introduction to move this book idea from a pro-
posal to a viable project. Also, of course, I am forever grateful for
the understanding support of my wife, Stephanie, who appreciates
the need to set time aside on nights and weekends for professional
matters. Without her outlook, I could not pursue this avocation.

SHIPHANDLING FUNDAMENTALS

FOR

LITTORAL COMBAT SHIPS AND THE NEW FRIGATES

CHAPTER ONE

Introduction

The U.S. Navy envisioned the littoral combat ship as a vessel that could respond quickly to operate in the near-shore environment. Its mission set included antisurface warfare, antisubmarine warfare, and mine countermeasures, but in addition to these missions, these ships were expected to conduct persistent surveillance and reconnaissance ahead of a main force. Once the Navy put these warships to sea, it discovered that their capabilities could be employed in an array of missions beyond the original intent, particularly given their comparative advantages in speed and maneuverability, as well as the ability to sail in shallow water. These advantages—speed, maneuverability, and shallow-water capability—are exclusively matters of shiphandling, so the shiphandler must know how to maximize their effects to make the most of these compact warships.

While traditional ships are fitted with propellers and rudders, this ship class is propelled by steerable waterjets that can be positioned up to thirty degrees to the left or right of centerline. This dynamic design gives the shiphandler precise control over the ship in both open and restricted waters. The full capabilities of the ship are realized once the shiphandler becomes comfortable with combining force vectors from each waterjet; the ship can then be made to perform truly extraordinary maneuvers, such as walking sideways.

The U.S. Navy's adoption of waterjets has fundamentally changed the calculus for shiphandlers, as it combines two key controllable forces, those once exerted by propellers and rudders, into one powerful force to control the ship. Chapter Four explains how the maritime concept of the "rudder" endures in these ships, but when conning officers order rudder changes, they are in fact ordering directional changes to propulsive thrust. The other controllable forces—tugs, anchors, and mooring lines—remain largely unchanged, as do the uncontrollable forces of wind, sea, and current. The task for the shiphandler is to gain mastery of these dynamic controllable forces to succeed in the age-old challenge of overcoming the uncontrollable natural forces.

While speed and agility are critical factors in defending the ship from attack, its maneuverability offers countless possibilities for accessing austere ports around the world. These ships have the ability to dock and undock without tug services; this fact, combined with a shallow-water capability, makes remote ports with underdeveloped tug services more accessible than ever. In this case, mastering the art and science of waterjet shiphandling will have a considerable strategic impact, as the ship's maneuverability directly affects the number of ports that are accessible. The nation's interests are often found in remote parts of the globe, and this ship, manned by skilled shiphandlers, can now reach those locations.

This chapter begins exploring LCS shiphandling by discussing the capabilities and limitations of these fast, light, and shallow-draft vessels. Understanding their characteristics will prove critical to the rest of this book, which will discuss in depth the relationship between the controllable and uncontrollable forces, waterjet management, and best practices for employing waterjets to exercise complete control over the ship.

FREEDOM- AND *INDEPENDENCE*-VARIANT CHARACTERISTICS

While the USS *Freedom* (LCS 1) and USS *Independence* (LCS 2) variants present starkly different designs, both sail fast in shallow waters. No matter the variant, the ship class's defining features are its shallow draft, high speed, and maneuverability.

The *Freedom* variant is based on a relatively traditional naval combatant design, with a single steel hull and a length and width commensurate with those of a corvette or frigate. It is 378 feet long with a fifty-seven-foot beam, and it displaces roughly three thousand metric tons when fully loaded. The ship draws about thirteen feet and can exceed forty knots. The *Independence* variant is based on the design of a commercial high-speed ferry, with a trimaran design that features an aluminum main hull and two aluminum outriggers, referred to as "amahs."[1] The ship is 419 feet long with a 104-foot beam, but like the *Freedom* variant it displaces roughly three thousand metric tons when fully loaded. The *Independence* variant draws about fourteen feet, and it too can exceed forty knots.[2]

Waterjet Propulsion

Both variants have engineering plants composed of two diesels and two gas turbines that drive four waterjets, as illustrated in figure 1-1. The *Independence* variant has separate drive trains for each

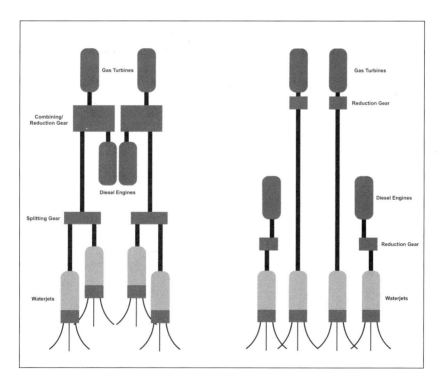

FIGURE 1-1   *Freedom*-Variant and *Independence*-Variant Drive Trains

waterjet; each engine has its own shaft, which drives a dedicated waterjet. The two diesel engines and waterjets are set outboard, and the two gas turbines and waterjets are set inboard. Rated at twelve thousand horsepower (HP) each, the two diesel engines together have sufficient power to drive the ship at more than eighteen knots; the 25,000 HP gas turbines are required for maximum speed. All four waterjets are steerable up to thirty degrees in either direction.

In contrast, as illustrated in figure 1-1, in the *Freedom* variant the diesels and gas turbines on each side output their power through a complex gearing system. Power from the port diesel and gas turbine power is combined into a single shaft on the port side, while the starboard diesel and gas-turbine power combines into a shaft on the starboard side. Each shaft passes through another gear that splits the power between the two waterjets on the respective side. The outboard waterjets are steerable up to thirty degrees in each direction, but the inboard waterjets are fixed and serve as boost jets that engage only at higher speeds. Which waterjets are selected by the control system depends on the desired thrust; the arrangement is governed by a system of plant *configuration states*. These states range from State 0 to State 9, as outlined in table 1-1. The diesel engines are rated at about 8,600 HP and together have sufficient power for more than twelve knots. The gas-turbine engines are much more powerful, rated at 48,000 HP, and their thrust is required to reach the ship's maximum speed.

Each waterjet in both variants controls longitudinal thrust by moving a reversing plate within the waterjet. As in controllable/reversible-pitch (CRP) systems on U.S. Navy cruisers and destroyers, the drive shaft always rotates in the same direction; but in contrast to adjusting propeller pitch in CRP systems, the resultant thrust is changed by altering the angle of the reversing plate. While the waterjet impeller thrusts water aft, the reversing plate can deflect the water forward under the hull, producing a resultant thrust vector astern. This plate can be adjusted to produce waterjet thrust from 100 percent astern to 100 percent ahead; intermediate positions produce thrust both forward and aft in combinations that yield resultant thrust vectors either ahead, astern, or stopped.

TABLE 1-1 *Freedom*-Variant Plant Configuration States

STATE	ENGINES ONLINE	WATERJETS ONLINE
0	None	None
1	Starboard diesel	Starboard steerable (outboard)
2	Port diesel	Port steerable (outboard)
3	Port and starboard diesel	Port and starboard steerable (outboard)
4	Port and starboard diesel	All waterjets
5	Port and starboard gas turbine	Port and starboard steerable (outboard)
6	Port diesel, starboard gas turbine	All waterjets
7	Port gas turbine, starboard diesel	All waterjets
8	Port and starboard gas turbine	All waterjets
9	All engines	All waterjets

All four engines can be controlled electronically from the bridge with *combinators*, which employ a single assembly to set both the thrust and direction of each waterjet. The thrust can be adjusted from a throttle setting of 1 to 10 ahead and from 1 to 10 astern, marked in whole-number increments, with 0 indicating "stop." The throttles move smoothly across the entire range, meaning that they do not snap between whole-number increments, so the throttles can be positioned between whole number settings. The combinator can also be set from thirty degrees left to thirty degrees right, the direction commensurate with the intended direction change of a ship driving straight ahead. In other words, a waterjet positioned left would be similar to left rudder and would ultimately cause the ship to turn to the left. However, as will be discussed in Chapter Two, in the section on waterjet basics, the "left and right" terminology is replaced with such terms as "toe-in" and "toe-out" during docking and undocking, to convey better the intertwining nature of these engines in waterjet management.

The physical construction and arrangement of the combinators is different in the *Independence* and *Freedom* variants, as shown in photo 1-1. On the *Independence* variant, there is only one set of four combinators, between the two watchstander chairs, each combinator specifically wired to one engine-waterjet drive train. The *Freedom* variant has two sets of two combinators, one set of combinators adjacent to the port chair and the other adjacent to the starboard chair. The *Freedom* variant, which routes engine output power through a combining gear, requires only two combinators: one for the port side and one for the starboard side. Additionally, in that variant only the outboard waterjets are maneuverable, so adjusting the combinator angle will affect only one waterjet on each side. Because the inboard jets are employed only at higher speeds, low-speed maneuvering evolutions like docking and undocking require only the outboard waterjets.

The *Independence* variant's additional two steerable waterjets provide even more maneuverability. This variant also offers a distinct shiphandling advantage by virtue of its bow-mounted thruster, shown in photo 1-2. This 360-degree bow thruster is capable of propelling the ship at speeds up to five knots, which represents a fallback option in the event of a casualty affecting all four drive trains. More routinely, its value to the shiphandler is that it provides excellent control over the bow during docking and undocking. A

PHOTO 1-1 *Freedom*-variant (left) and *Independence*-variant (right) combinators
    Photo by author

Photo 1-2 *Independence*-variant thruster
Photo by author

bow thruster makes it possible to dock and undock easily without the assistance of tugs, even in environmental conditions as challenging as a twenty-knot opposing wind. In other words, using the thruster the ship can walk laterally to the pier against a twenty-knot offsetting wind or walk laterally away from it against a twenty-knot onsetting wind.

The thruster is also available for shiphandling in situations beyond orthodox docking and undocking evolutions. It can be deployed at any time the ship is making less than five knots through the water, and it can remain deployed at speeds up to ten knots. As will be discussed in detail in Chapter Seven, on special evolutions, the thruster can be employed creatively during such tasks as small-boat operations and mission-package deployment.

Large Sail Area, Shallow Draft

The ship is specifically intended to patrol in water previously considered too shallow for U.S. Navy warships. With their thirteen-foot and fourteen-foot drafts, respectively, the ships of the *Freedom* and *Independence* variants are able to enter waters previously considered out of reach. This design characteristic, however, has implications for the shiphandler that are particularly important when entering and leaving port.

As will be discussed in Chapter Two on controllable and uncontrollable forces, the uncontrollable forces on the ship have not changed: wind, current, and sea state. For any other variant of ship, the effects of current are considered to be far more impactful than those of wind, because water is a denser medium than air. The large volume of water moving beneath the surface grabs the hull and pushes it in the direction in which the current is flooding or ebbing. In fact, a common rule of thumb for cruisers and destroyers is to consider current thirty times more powerful than wind—that is, one knot of current is equivalent to thirty knots of wind. This rule of thumb, however, does not fit the littoral combat ship, whose shallow draft yields a radically different ratio between the surface area above and below the waterline. For this ship class, one knot of current is roughly equivalent to fifteen knots of wind.[3]

ART VERSUS SCIENCE OF WATERJET SHIPHANDLING

When novice shiphandlers are first exposed to shiphandling, they are given basic rules of thumb like the one just mentioned for maneuvering the ship. For example, a conning officer in a propeller-driven ship might be taught that in order to twist the ship to starboard the port engine must be ordered ahead one-third, the starboard engine back one-third, and the rudder to starboard. This prescriptive approach is enough to get novices started, but in time they will learn to watch how the ship reacts to engine orders and the environment. Perhaps their ship's one-third backing bell is stronger than its one-third ahead bell, requiring the shiphandler to compensate with a two-thirds ahead bell. Maybe their ship's topside configuration makes it more difficult to twist into the wind than with the wind, which would demand, in the former case, a two-thirds or standard bell. When shiphandlers begin to notice these cues, they are transitioning from novice to advanced shiphandling; the difference becomes evident in their ability to let go of strict rules of shiphandling in favor of a more artful approach.

This effect is even more pronounced in waterjet shiphandling, because of the vast number of combinations available to maneuver the ship. Whereas propeller-driven ships have fixed propulsion trains that direct thrust either forward or aft, the steerable waterjets can exert propulsion thrust up to thirty degrees left or right. This flexibility offers an array of combinations that achieve similar maneuvering effects. One who gains command of force vectors from steerable waterjets and grasps the vast range of combinations that achieve the same effects is one who has mastered both the art and science of waterjet shiphandling.

Employing both the art and science of shiphandling can be a challenging task, particularly in circumstances where the two approaches contrast in nature. It can be especially difficult for individuals who have either very strong or very weak mathematical abilities. Those with mathematical minds may view shiphandling in stark terms of vector math, whereas those without these skills may prefer to feel their way through shiphandling evolutions. Fanatical allegiance to either camp can be perilous, as die-hard mathematicians may become so wedded to their equations that they fail to anticipate the unexpected, and the math-averse can find themselves hunting for the right combination even in fairly straightforward situations. In general, it is best not to adhere strongly to one view or the other but instead to develop a combination of abilities for various situations. Shiphandling crises often develop very quickly, and even the most skilled shiphandlers cannot predict what tools they will need to conquer them.

WATERJET VERSATILITY: BOTH AN ASSET AND LIABILITY

The versatility of steerable waterjets enables the shiphandler to be more creative in maneuvering the ship than with propeller-driven ships, but this kind of ship driving can be troublesome to a professional military force that is built on a culture of *simple, repeatable* procedures. A shiphandling crisis may not be the time for creative thinking or for experimentation with something new. The Navy has succeeded throughout the years by learning from past experiences, developing emergency procedures for many situations, and training to employ them long before they are needed.

Naval officers drive ships in environments that are a balance between operations and training—that is, in which junior officers learn shiphandling while also accomplishing operational objectives—and the priority between the two shifts according to the situation. For example, during a quiet day of man-overboard training in the open ocean off southern California, the junior shiphandler is given room to make mistakes, as the ship's captain instructs on how to handle the ship better. In contrast, a busy day of operations overseas, such as close-quarters maneuvering alongside an oiler, does not afford much room for error. The people involved are the same—the novice shiphandler conning the ship and the commanding officer providing guidance—but the balance between training and operations shifts, such that role of the captain is more directive in nature. In the less intense environment, novice shiphandlers learn by trial and error, but in more constrained situations they learn by successfully completing events under close supervision.

As one can imagine, when the environment is more operational than instructional and the level of risk is elevated, there must be a clear understanding between the novice shiphandler and the captain as to what is expected. In these circumstances, the versatility of steerable waterjets can be problematic. The shiphandler may have vast combinations of thrust direction to choose from, but unless the options are narrowed to a subset that will produce expected outcomes, the commanding officer is relegated to an observer, and the captain's ability to identify inappropriate actions and intercede depends on his or her quickness at vector math. Close-quarters maneuvering does not afford much time for reaction, let alone time for correcting improper action. The safer and more efficient approach is to establish a shared method linking observations to shiphandling actions.

This book aims to do just that, by providing *simple, repeatable methods.* Using them the shiphandler and commanding officer will share an understanding of how to control the ship, enabling them to take action quickly in response to particular ship motions in crisis situations. Training to these methods is critical to being ready when an emergency develops.

The chapters that follow aim to achieve two objectives. First, the shiphandler must understand the full range of combinations available on ships equipped with steerable waterjets. Then, these combinations will be narrowed to a set of *simple, repeatable methods* that facilitate a clear understanding between the junior shiphandler and the captain.

This book is structured in nine chapters that cover the principles and practical applications of waterjet shiphandling for this ship class. Building on this first chapter's discussion of basic shiphandling characteristics, the next chapter covers the controllable and uncontrollable forces in shiphandling. The uncontrollable forces are as old as time, but Chapter Two will consider how the ship's specific equipment can be used to control vessel movements despite them. The discussion continues in Chapter Three with a deep exploration of waterjet vector management, demonstrating how the possible countless thrust-vector combinations provide a vast array of shiphandler options. It will explore practical applications that achieve typical ship maneuvers, such as twisting in place and laterally walking the ship. This discussion will focus on simplifying an inherently complex challenge by applying methods proven at sea, and it will present standard commands that have been tested on the littoral combat ship. As the discussion develops, this book will systematically apply these methods to common shiphandling tasks, including pierwork, channel driving, underway replenishment, anchoring, mission-package operations, small-boat launch and recovery, and heavy-weather maneuvering. The book concludes with a chapter on maximizing use of the training facility to prepare for at-sea shiphandling.

CHAPTER TWO

Controllable and Uncontrollable Forces in Shiphandling

The forces that act on a ship can be divided into two categories: uncontrollable and controllable. The uncontrollable forces are those colloquially described as the powers of Mother Nature. The controllable forces include all man-made factors that allow the shiphandler to control the ship. Resist considering uncontrollable forces as evil and controllable forces as good—the earth sometimes pushes the ship in a favorable direction. The critical difference between the two is that the shiphandler can choose the direction in which controllable forces act. As for the uncontrollable forces, the shiphandler must know the direction they are coming from and how they will affect the ship.

One of the perpetual tasks of shiphandlers is figuring out *why* the ship is moving in a certain way; the most advanced shiphandlers concentrate on observing *how* the ship moves and determining *what* forces are causing it to do so. The only difference between the advanced and novice shiphandler is that the advanced mariner can more quickly recognize a pressure differential and apply counteracting pressure by means of a controllable force.

This chapter begins with the uncontrollable forces, to understand how the ship would be affected if it were subject to natural influences alone. The uncontrollable forces are what they have always been and should be familiar to experienced shiphandlers, but they are presented here for the sake of novice shiphandlers, particularly those whose littoral combat ship is in fact their first ship. Then this chapter will review the controllable forces at the shiphandler's disposal. The practical application of these forces will be detailed in subsequent chapters, which are divided by types of shiphandling: pierwork, channel driving, special evolutions, and the open ocean. Between uncontrollable and controllable forces, however, it makes little difference which shiphandlers learn first, as long as they eventually understand the impact and practical applications of each.

UNCONTROLLABLE FORCES

As mentioned in the introductory chapter, some facets of shiphandling are timeless, and the uncontrollable forces fall into this category. From the first days of sail, mariners have never been able to control wind, current, or sea state. The advanced shiphandler learns through study and experience to understand these forces affecting the ship, to anticipate ship movements resulting from these forces, and to apply the controllable forces available to correct undesired ship movements. It is important to note that not all movements resulting from uncontrollable forces are necessarily hazardous; the most skilled shiphandlers maneuver the ship so as to make these forces work for them.

Wind

Of the three uncontrollable forces, wind most frequently affects ships. Current is most apparent in confined harbors where the topography funnels tidal flow and increases its effect, but it is less influential in the open ocean, where the effect is spread across a larger area. Sea state, in contrast, is most impactful in the open ocean, where there are no obstructions to dampen its effect, but it is felt inside protected harbors only in particularly adverse weather. Wind, however, is present every day, whether in port or at sea. This may make wind seem like a perpetual nuisance, but incorporating it into the shiphandling plan can allow the conning officer to maneuver the ship more effectively.

To understand how wind affects the ship, consider the ship's movements if it is pushed only by the wind, with no other forces at play. If the wind were coming from broad on the bow—045 or 315 degrees relative—it would push the bow away until it was applying pressure equally on the bow, beam, and stern, that is, with the ship falling beam to the wind. Wind coming from the quarter—135 or 225 degrees relative—would push the stern away until it was applying equal pressure to the stern, beam, and bow, again bringing the ship beam to the wind. The logical conclusion from these scenarios is that if left to twist, literally, in the wind, the ship will fall beam to the wind. Even if the ship is pointed directly into or away from it, the perpetually shifting wind will eventually catch the port or starboard side and cause the ship to pivot until it is beam to the wind.

In reality, however, this does not mean that the ship will always settle with the wind directly on the beam, on a relative bearing of 270 or 090. Differences between the sail area forward and aft will cause the bow or stern to be more affected than the other. For example, the *Independence* variant's large sail area aft will result in the stern moving faster in response to wind than the bow. Still, the ship will settle with the wind roughly on the beam, so parsing the difference in force vectors between the bow and stern has little value when actively handling the ship. It will suffice to conclude that the stern will move more quickly but that the ship will eventually fall roughly beam to the wind.

It is also important to understand that this movement is not linear. Crenshaw notes that the wind's force is proportional to the square of its velocity and varies with the ship's sail area facing the wind and the form of the superstructure. Consequently, as the ship pivots and presents more surface area, the resultant force is increased. This is important to remember when trying to maneuver out of a beam wind; more power will be required to pivot the ship with the wind near the beam than when it is broad on the bow or on the quarter. One other effect related to this nonlinear relationship is that doubling the wind speed quadruples the force on the ship.[1]

Wind has a particularly great effect on the ship class due to its light weight and shallow draft. First, since the ship is very light, designed so as to maximize the thrust-to-weight ratio, the wind is able to push it with relative ease. Even fully loaded, the ship displaces little more than three thousand tons, about a third the weight of an *Arleigh Burke*–class destroyer. Second, wind is trying to push the ship against the resistance of the water on the leeward side. The ship's shallow draft presents less area against which this resistance can act to hold it in place, so a relatively low-velocity wind will move this class of ship more quickly than it will a deeper-draft vessel.

Current

It is generally accepted that current has a much larger effect on the ship than wind, because water is denser than air. From the above discussion on wind, one might presume that the ship's shallow draft would lessen the effect of current on it, but that is only relative to the wind's effect. The ship's lighter weight also makes it susceptible to the power of the flow of water; it will take a less strong current to move this ship at the same rate as a heavier ship. As a rule of thumb, remember, one knot of current is equal to fifteen knots of wind for this ship class.[2]

Similar to wind, current will carry the ship in the same direction as its flow. The difference between wind and current, however, is that in protected waterways current direction and velocity can change radically. Twists and turns in the topography can direct the current in multiple directions, and the velocity can change near stationary objects. Furthermore, owing to the nature of ebb and flow, within minutes the current will flow in one direction, then stop, and then flow in the opposite direction. To the novice shiphandler, current may seem like an erratic hazard, but current is fairly accurately predicted. The advanced shiphandler knows how to incorporate this uncontrollable force into the shiphandling plan to maneuver the ship more effectively.

The most dangerous circumstance in confined waters is to have the current on the beam. First, the current's effect on the ship, like that of the wind, is not linear. It is proportional to the square of the current velocity and varies with the underwater area facing the flow. The more underwater area presented by the ship, the more lateral force required to hold the ship on course.[3] Second, the ship's most powerful controllable force—its propulsion—exerts a longitudinal force exactly perpendicular to the beam. In this circumstance, when the shiphandler requires the greatest control over the ship, the strongest controllable force is least effective. This factor is ameliorated by the littoral combat ship's steerable waterjets, but the disadvantage remains principally the same. In the worst case, in the absence of controllable forces, the ship will be carried along with increasing speed until the current can no longer accelerate the ship against the water's resistance.

Ocean currents present similar challenges, but their direction and speed are largely constant over a given area of water. Shiphandlers often overlook published predictions for these forces, because the open ocean presents less risk of grounding, collision, or allision (that is, striking a stationary object), but the advanced shiphandler knows how to use these predicted forces to the ship's advantage. For example, plotting a course to take advantage of a following ocean current can save both time and fuel. Shifting a track by as little as ten miles can be the difference between a following current and neutral water.

Seas

"Seas" refers to surface ocean effects on the ship. Seas influence the ship in the form of surface waves that are generated by local winds or that appear as long swells produced by winds farther away. Whereas ocean currents will consistently push the ship in a certain direction, seas will affect how the ship pitches, rolls, and yaws. In addition, seas will affect how the ship responds to certain maneuvers. The shiphandler must understand the effect of seas on the ship, because unless these effects are anticipated and counteracted, the crew and the ship itself may find themselves in great danger.

Nearly all seas are caused by wind, but the appearance of these seas when arriving at the ship varies depending on the distance traveled and how long the winds have persisted. Seas created by local winds exhibit ripples or caps that proceed in the same direction as the wind. These visual cues dissipate when the wind dies down, though the waves beneath them continue moving in the same direction. The absence of obstructing forces allows these waves to continue for thousands of miles, often ending their journey only when reaching shore.[4]

The distance that the wind can blow unobstructed in the same direction is called *fetch*; the longer the fetch, the more significant the seas that will be generated.[5] In fact, the most dangerous seas are the result of *unlimited fetch*, built up by wind unobstructed over great distances. While geography often imposes at least some limits on wind, in some areas of the ocean seas can build over hundreds or thousands of miles. Once a ship leaves the protected harbor, it may immediately face dangerous seas caused by unobstructed wind from the ocean.

Waves can travel such long distances that they often enter areas where the local wind may blow in a different direction, creating its own surface effects. Waves generated by storm winds in a distant area may have so much volume and momentum that they are not affected by opposing local winds. For example, seas that originate in a storm a thousand miles away at 270 degrees true from a ship may arrive where local winds from 180 degrees true are creating smaller seas also impacting the ship. Some mariners distinguish between the two by applying different names—for example, describing waves with ripples or caps as *seas* and those without as *swells*.[6] In the end, both are seas that impart uncontrollable forces on the ship. But swells can be more dangerous for this ship class, because the extraordinary volume and momentum presented by these seas can be challenging for a vessel so light.

When faced with seas from different directions, the shiphandler must identify each type and determine which are the *predominant seas*. For example, if the shiphandler observes short-period swells from the west but a thirty-knot surface wind is blowing from the north, the predominant seas would be from the north. Conversely, if rolling, long-period swells from the west move into an area with a ten-knot surface wind from the north, the predominant seas will be from the west. In cases where the seas are from different directions and neither is predominant, they are referred to as *confused seas.*

Chapter Eight, on heavy-weather shiphandling, will discuss seakeeping strategies in depth, but even in less extreme conditions the shiphandler must always keep the seas in mind. This ship class rides high in the water; and although it is able to sail the open ocean, its design is optimized for the shallow waters of the nearshore environment. Excessive ship movement at sea—pitch, roll, yaw, and heave—can have deleterious effects on the crew, in the form of fatigue, injury, and even illness. The least incapacitating of these consequences, fatigue, may in fact become the most dangerous over time as watchstanders who are prevented from resting must continue to stand their watches, subsequently endangering the ship. The most incapacitating consequences, injury and illness, considerably affect the minimally manned crews, which do not offer the luxury of replacements should the primary watchstanders become incapacitated. The loss of able watchstanders increases the on-watch hours of the others, worsening their fatigue and further hazarding the ship.

Because of these factors, the shiphandler must live by the golden rule of seakeeping: *always know where the seas are coming from!* Conning officers may not be able to stop the ship's motion at sea completely, but they cannot begin to alleviate the problem if they do not understand the forces acting on the ship.

THE NATURE OF WATER

While studying hydrodynamics is not a prerequisite to become a good ship driver, the journey from novice to advanced shiphandler should include a growing appreciation for the nature of water, the principal medium that affects ship movements. Understanding the effect of moving water on the ship will allow the shiphandler to anticipate movement and take early action with the controllable forces to counteract undesired effects.

To begin with, the shiphandler must understand the first principle of water: *the flow of water is the result of pressure differences.* Water, unlike gas, is incompressible. Any force applied to water causes a pressure differential between water in one area compared to that in another, which causes water to flow from higher-pressure to lower-pressure areas. The practical lesson is that all ship movements caused by water can be explained by pressure differential.[7] By understanding the nature of water, the shiphandler can quickly determine the pressure difference that is causing ship movements. Two key concepts regarding the nature of water are Bernoulli's principle and Venturi forces.

Bernoulli's principle states that in a fluid with uniform flow, the sum of the following will remain constant: energy from velocity, energy from pressure, and potential energy from elevation.[8] Since the elevation at sea level remains unchanged, this formula narrows down to a function of velocity and pressure. The water in the ocean is subject to only two components of pressure: static pressure from the weight of the water and dynamic pressure caused by the movement of surrounding water. Changes in depth at sea are not significant enough to change the static pressure of the water on a surface ship, so we consider the static pressure to be constant. Therefore, the pressure of water in the ocean is affected only by changes in velocity.[9]

Consider how water must travel from the bow to the stern as the ship passes through it. Imagine a line of water particles on the surface one hundred yards across the path of an oncoming ship. At first, they are all traveling the same speed, which is the speed of the ocean current. They are in alignment before the ship arrives, and they will be in alignment again—nominally, of course—after the ship passes through them. The ship is simply cutting through these particles of water. The particles to the left and right of the ship's path will not be touched, so they will only travel as fast as the ocean current carries them. The particles in the ship's path, however, will be pushed to port or starboard, pass down the side of the ship, and meet up again astern. The water particles in the ship's path are traveling a greater distance than the untouched particles, yet they all arrive astern at the same time; so the water passing down the ship's sides *must be traveling faster*.[10]

This observation leads directly to Venturi forces. Since Bernoulli's principle states that the sum of velocity and pressure remains constant, increases in the water's velocity cause decreases in its pressure. As the water moves down the ship's side and its velocity increases, an area of low pressure is created along the hull. This is particularly important when maneuvering in close proximity to other objects, as the low-pressure area created can appear to draw the ship toward danger. Venturi forces are most dangerous when two ships are steaming alongside one another, such as during an underway replenishment, because both ships create low-pressure areas along their hulls. The combination of these low pressures causes the ships to seem to be drawn together. In reality, it is the flow of higher-pressure water from the outboard side of each ship into this low-pressure area that pushes the ships together, and each ship must apply counter-rudder to combat these forces. The lightweight design and shallow draft of the littoral combat ship offer reduced resistance to Venturi forces that push ships together. This phenomenon will be discussed in more detail in Chapter Seven, on special evolutions.

CONTROLLABLE FORCES

To overcome these uncontrollable forces, shiphandlers must understand what forces are at their disposal and constantly study their capabilities and limitations. Since the Navy transitioned from sail to power, the controllable forces have been characterized as propulsion, rudders, tugs, mooring lines, and anchors. The Navy's adoption of waterjet propulsion in this ship class has changed this characterization somewhat.

Waterjets serve as both propulsion and steering, eliminating the requirement for rudders to maneuver the ship. The only difference between the *Independence* and *Freedom* variants, as discussed in Chapter One, is that the *Independence* variant has a 360-degree bow thruster. Therefore, for this ship class, the controllable forces are waterjets, bow thruster (*Independence* variant only), tugs, mooring lines, and anchor. For clarity, the *Independence* variant does in fact have two rudders, part of the ride control system, but these rudders are employed for steering only at higher speeds, to reduce excessive stern movement caused by steerable waterjets; they will not be considered in this discussion, which focuses largely on lower-speed shiphandling.

Waterjets

Propulsion traditionally transmits a force longitudinally along the axis of the keel to move the ship ahead or astern. The waterjets serve the same purpose as traditional propellers, to translate power from the ship's engines into a force that moves the ship through the water. The ability to swing the waterjets left and right, however, has changed how we think about maneuvering the ship.

The previous chapter described how waterjets control the amount of longitudinal thrust by moving a reversing plate. As in the controllable/reversible-pitch system of Navy cruisers and destroyers, the drive shaft always rotates in the same direction, turning the impeller that pushes water through the waterjet; the reversing plate changes the direction of the thrust by altering the output path of the water. To make the ship move aft, the reversing plate deflects the water forward and under the hull, producing a resultant force aft. As noted in the previous chapter, this plate can be adjusted to produce waterjet thrust from 100 percent astern to 100 percent ahead, and intermediate positions produce thrust both forward and aft that combine to yield a resultant force that is either ahead, astern, or stopped. As we have seen, the main difference between the two variants is that all four waterjets on the *Independence* variant are steerable, whereas the *Freedom* variant steers with the outboard waterjets only.

Because these waterjets can be steered thirty degrees to port or starboard, the engines can provide propulsion forward, aft, left, and right. Chapter Three will discuss waterjet vector management in depth, but at this point it is sufficient to understand that changing the direction of waterjet thrust relative to the keel will produce varying thrust vectors both longitudinally and laterally. Unless the reversing plate is perfectly positioned to route equal amounts of water both forward and aft, each waterjet always produces longitudinal thrust. Similarly, unless they are perfectly centered, to point directly aft, each steerable waterjet exerts a lateral force moving the stern left or right.[11]

In reality, it is somewhat simplistic to summarize waterjets as both a propeller and rudder, because rudders are only effective when water is flowing across them. Rudders are optimally positioned directly astern of propellers to maximize the flow across their surfaces, creating hydrodynamic lift in the horizontal plane that moves the stern left or right. It is possible to steer when driving a ruddered ship astern, but rudders are far less effective astern, as the ship must have enough speed for the rudders to produce lift without propeller wash across them. Fortunately, waterjets are effective both ahead and astern, because they do not need to create lift to move the stern. In fact, the shiphandler can steer the ship astern at speeds as low as one-tenth of a knot.

Waterjets have effectively eliminated the hazard of lost steerageway. When insufficient water is moving across a rudder's surface to produce lift, the ship is considered to have lost steerageway; the stern is no longer controllable. Waterjets, however, are always discharging water, constantly maintaining control over the stern. Even when the ship is dead in the water, the shiphandler can arrange the waterjets to counter each other to maintain control over the stern. While ruddered ships sometimes have to choose between stopping and maintaining steerageway, waterjet ships can do both simultaneously.

Like a rudder, however, the waterjet turns the ship by moving the stern away from the intended direction of travel. When the shiphandler wants the ship to turn right, moving the waterjets to the right will propel water out toward the starboard quarter. This force results in an opposite reaction, moving the stern to port, which pivots the ship's bow to starboard. The turn is facilitated by high-pressure areas that emerge on the port bow, as the ship's port side meets the resistance of the ocean.[12] Ruddered ships make their tightest turns from a dead stop, since the full force of the propeller is translated to the rudder, and the ship does not slide away from the turn because of forward motion.[13] This effect is amplified with waterjets, because 100 percent of their thrust moves in the direction of the turn.

When turning the ship at higher speeds, the shiphandler must keep in mind that thrust is directed away from the ship's axis. For the *Independence* variant, maneuvering all four waterjets at forty knots will yield a sharp turn, but the forward thrust that had been propelling the ship ahead is now projected off axis, and the ship's longitudinal speed drops quickly. The *Freedom* variant is less susceptible to this effect, since only two of its four waterjets are steerable, and the *Independence* variant can ameliorate it by only using one or two waterjets to steer at high speeds. Similarly, the *Freedom* variant can maintain maximum speed in a turn by steering with only one waterjet.

Highlighting another advantage of waterjets in this ship class, it should be noted that waterjets are not subject to the side thrust that affects propellers. The impellers that drive the water are contained within the waterjet assembly, and the spacing between the outer edge of the impeller and its shroud is minimal. Consequently, even though all four waterjets turn in the same clockwise direction, the stern is not subjected to these lateral forces.

The ship's light weight is a critical factor when employing waterjets during pierwork. The ship's high speeds are made possible by its extraordinarily high thrust-to-weight ratio. An *Arleigh Burke*–class destroyer is 30 percent longer and 260 percent heavier than the *Freedom* variant, yet the latter has almost 15 percent more thrust through its waterjets.[14] The thrust-to-weight ratio is an enormous warfighting advantage in open water, but alongside the pier it can quickly spell disaster if not carefully controlled. Destroyer captains rarely use more than a two-thirds bell alongside the pier. On the littoral combat ship, however, maneuvering adjacent to the pier rarely requires more than two-tenths of the maximum throttle. The application of this force will be discussed further in Chapter Five, on pierwork, but it is important to note at this point that from a thrust-to-weight perspective the ship is so powerful that the waterjets must be employed judiciously when alongside the pier. The shiphandler's task is to determine how much power is necessary to move the ship and then avoid using more.

Bow Thruster (*Independence* Variant Only)

Any vessel with a bow thruster possesses a distinct shiphandling advantage when maneuvering alongside the pier, because the thruster can control the bow without the assistance of tugs. It should be noted, however, that bow thrusters do not generate nearly as much power as modern tugs, so their ability to overcome uncontrollable forces is far more limited.

Still, a combination of the thruster and waterjets allows the shiphandler to shift the pivot point to any position between the bow and the stern, and more incredibly, without increasing speed forward or aft. The bow thruster is maneuverable throughout 360 degrees and, as noted previously, is capable of propelling the ship up to five knots on its own. It provides a balance of power and precision that makes it possible to dock and undock the ship without tugs in up to twenty knots of lateral wind. For example, again as noted above, with the use of the thruster the ship can walk sideways to the pier against a twenty-knot offsetting wind, and it can walk away from the pier with a twenty-knot onsetting wind.

The thruster is also used in shiphandling evolutions other than docking and undocking. It can be lowered any time the ship is making less than five knots through the water, and it can remain deployed up to ten knots. As mentioned above and as will be discussed in detail in Chapter Seven, the thruster can be employed creatively during such shiphandling evolutions as small-boat operations and mission-package deployment and recovery. As useful as this tool is for shiphandling, however, its primary purpose is as an emergency option in the event of a casualty to all four drive trains.

Tugs

The powerful tugs that are available today offer significant force in order to control the ship laterally while alongside the pier. The resultant movement of the ship is a function of power and resistance as with the other controllable forces, so the ship's lightweight design and low underwater resistance make today's extraordinarily powerful tugs very effective in handling these ships in various environmental conditions.

In and around Navy bases, particularly within the United States, tugs have become both more powerful and more controllable in recent years. The Navy's *Z-drive tugs* have become ubiquitous in fleet concentration areas. The power inherent in these tugs is critical for safely docking larger ships, but their power is invaluable in controlling smaller ships in weather that may have previously precluded them from entering port. Z-drive tugs fall under the category of *directed thrust*—that is, the tug can aim its propeller in any direction to change the resultant thrust. With this 360-degree maneuverability, the tug does not have to maneuver relative to the ship to gain leverage. The tug can maintain the same relative position and still pull or push the ship in any direction, which also allows it to exert the same amount of power toward and away from the ship.[15]

Single- and twin-screw tugs, however, have greater limitations. First, the backing power of these tugs is less effective than their ahead thrust. As a rule of thumb, a conventionally driven tug when backing exerts about two-thirds the power that it can ahead.[16] Second, without the ability to direct thrust 360-degrees, these tugs must maneuver along the axis of the required resultant thrust on the ship. For example, if a tug on the bow is called to pull the ship to port but the line from the ship to the tug tends 225 degrees relative from the ship, the tug must reposition itself so that the line tends 270 degrees relative before pulling. Otherwise, it would pull the ship to the left and aft.

This limitation is particularly important when the ship is making slight way ahead or astern. The tug captain must constantly reposition to be ready to answer a call for power either toward or away, and if the ship is making greater than two knots the tug will have difficulty remaining in position and will require additional reaction time to answer a call for power. If the ship is moving faster than three knots, the tug will not be able to maintain its ninety-degree angle off the ship, and its pulling and pushing power will be significantly reduced. It should be noted that a single-screw tug is even more challenged to remain alongside, because it does not have the twisting ability of a twin-screw tug to keep its stern in place. Finally, these tugs may need to lean against the ship to pivot into position.[17] As explained below, even this small amount of pressure against the hull can have considerable effects on the pivot point of a ship that displaces only three thousand tons and has very little resistance below the waterline.

This ship class is so maneuverable with the steerable waterjets on both variants, particularly with the bow thruster on the *Independence* variant, that one might wonder why tugs would even be employed. But even when tugs are not required to dock the ship, they still serve as a safety net in most pierwork, even in the most benign conditions. The prudent shiphandler will tie on at least one tug and order it to "keep a slack line," even if only to release it after the evolution without ever requiring its assistance. While any serious shiphandler appreciates the challenge of landing the ship without outside assistance, if an emergency or mishap develops it may be too late to call the tug to put over a line.

This approach to tug employment, however, is based on certain assumptions about the tug captain's abilities, which are sometimes less than certain. Several factors might convince a commanding officer in specific situations that tugs represent a greater hazard than maneuvering alone. Tug captains in austere ports overseas are often inexperienced in handling U.S. Navy ships, and effective communication of maneuvering orders to the tugs can be complicated by insufficient communications equipment or marginal proficiency in English. The commanding officer must also weigh the tug captain's experience in maneuvering this ship class, whose profile above the waterline makes it appear large and heavy but is actually light and responsive, because of its shallow draft and lightweight structure. This visually deceptive design may cause an inexperienced tug captain to provide too much power, a condition that can be more dangerous than too little power. Further, tug handling alongside aluminum hulls is far different than with stronger hulls, such as steel. A tug making aggressive contact with the aluminum hull of a ship of the *Independence* variant might cause more damage than incidental contact with the pier if the ship had docked without tug assistance.

The maneuverability of this ship class can also be disrupted by even the most capable and experienced tug captains. The light weight of the ship and its low underwater resistance make it susceptible to slight contact with an adjacent tug that has been instructed to remain clear. It is common practice for a tug captain expecting an order to push against the ship to rest the tug lightly against the hull to facilitate a quicker response. This anticipative contact allows the tug captain to provide immediate power without worrying about approaching the ship too aggressively. This type of contact has a negligible effect on larger, heavier ships, but it can significantly affect this lightweight ship class. For example, a tug resting against the bow can prevent the shiphandler from moving the bow, acting as an anchoring force and shifting the pivot point forward to the point of contact. Conversely, a tug could exert just enough counterforce on the bow to shift unexpectedly the pivot point aft. This type of contact is well intended, but these unanticipated forces on the ship can introduce confusion during pierwork, causing the shiphandler to make decisions based on incomplete information.

Mooring Lines

The novice shiphandler views mooring lines as merely a means to keep the ship tied to the pier. During pierwork, however, these lines are an effective way both to keep the ship from moving toward a hazard and to pull it in a desired direction. Figure 2-1 shows the nominal line configurations for this ship class.

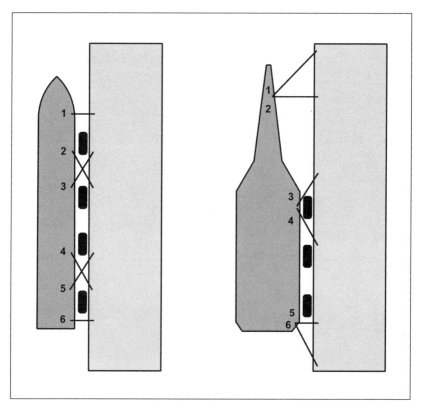

FIGURE 2-1 Notional Mooring Line Configurations for the *Freedom* Variant and *Independence* Variant

To understand how mooring lines can be used in shiphandling, look again at figure 2-1 and consider how the ship would be prevented from moving in a certain direction if only one of those lines was connected. For example, if all lines were taken in except line four—the after spring line on either variant—the ship could move in any direction except forward. Leaving this line singled up would be useful if there was a hazard in front of the ship, such as a quay wall or another ship. The shiphandler could leave it on until the ship had slight sternway and then take it in. In another example, consider how the ship would move if there was an offsetting wind and only a single breast line on the bow was left singled. The wind would blow the stern off the pier, but the bow would remain in

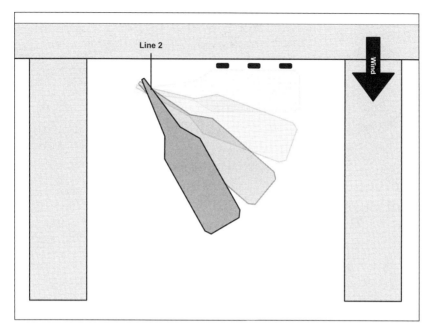

FIGURE 2-2   Using Line Two (Breast Line) to Undock the Ship from a Quay
Wall

place, allowing the ship to pivot on the bow. Figure 2-2 illustrates how this approach would be useful if, getting under way from a quay wall between two piers, where the shiphandler wanted to back out of the slip.

There is an important design factor for the *Independence* variant that must be considered when positioning lines during docking. The ship's side door, which is used for loading and unloading mission-package equipment, is located on the starboard side beneath the flight deck. Since this door must be lowered for loading and unloading operations in port, the shiphandler must ensure that its path is clear of lines during final positioning of the ship alongside the pier.

The Anchor

This ship class is outfitted with one anchor, which, in addition to holding the ship in place at anchorage, can prove useful in assisting the shiphandler in maneuvering the ship in tight quarters. Similar to a standard Navy Mark 2 lightweight anchor, it is designed in such a way that the flukes deploy and dig in to the ocean floor; it can be set only if pulled horizontally along the ocean floor. Sufficient chain must be paid out before any considerable strain is placed on the anchor to ensure that it will be pulled back and not up.[18]

The *Independence* variant uses cable instead of chain; more accurately, the ship has one shot of chain, which connects to a spool of cable. To be sure, the cable weighs less than chain of an equal length, which is precisely the reason that cable is used instead of chain. With other characteristics of the ship class, lightweight construction is a critical enabler of its shallow draft and high speed. One might conclude that the lighter cable would translate to decreased holding power while at anchor, but the weight loss should be considered relative to the weight of the ship itself. Since the cable is about one-third the weight of a standard chain, and the ship is about one-third the weight a similar-sized destroyer, the ratio remains roughly the same. Therefore, the traditional rule of thumb of paying out a length of cable five to seven times the depth of the water is a sufficient starting point for any anchorage. Just as the term "rudder" is maintained for these ships, "chain" can still be used during anchoring evolutions, regardless of variant.

During transits in restricted waters, making the anchor ready to let go in the event of an emergency is prudent. Should the ship lose all controllable forces and begin drifting toward shoal water, the anchor may prevent the ship from running aground. In this situation, the shiphandler must choose between two possible options to stop the ship. First, the anchor can be deployed at short stay, with just enough cable scope to allow the anchor to contact the bottom. The upward angle of the chain will prevent the flukes from digging in, but dragging the anchor along the bottom might provide enough resistance to slow the ship's movement toward danger.[19] Second, the shiphandler can pay out enough chain to provide a horizontal pull on the anchor, allowing the flukes to dig in. Doing so requires enough distance from shoal water to pay out sufficient chain, and trying to set the anchor at uncontrolled speeds greater than five knots risks damage to the anchor-windlass system that could result in complete loss of the anchor.[20]

The anchor can also be used creatively during shiphandling maneuvers to exert control over the bow without a tug. For example, as illustrated in figure 2-3, the anchor can be used to brake the bow against an onsetting wind. As the wind pushes the ship onto the pier, the anchor can be dropped approximately one hundred yards perpendicular from the pier. Slowly paying out the anchor chain under tension will allow the shiphandler to ease the bow toward the pier; once moored, the chain can be slacked to the bottom to reduce the risk of obstructing a passing vessel. During undocking, having left the anchor in this position, the shiphandler can heave around on the anchor to pull the bow slowly away from

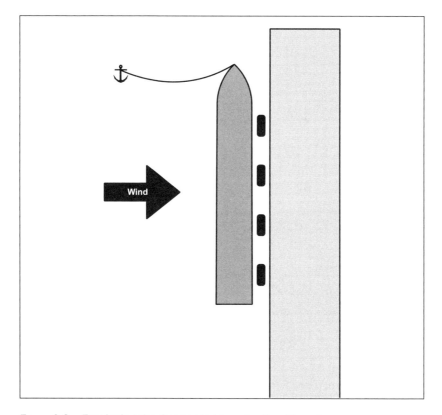

FIGURE 2-3   Employing the Anchor in Lieu of a Bow Thruster

the pier.[21] This tool is particularly useful for the *Freedom* variant, which does not have a bow thruster, but it is also useful for the *Independence* variant when the thruster is not available or the water is so shallow that the thruster cannot be lowered. Looking at the anchor as a means of controlling the bow opens a world of possibilities that are essential when operating in austere ports around the world.

Since this ship class only has one anchor, the limitations are worth mentioning. First, the ship is unable to conduct a standard Mediterranean moor, where the stern is tied to the pier with mooring lines and the bow is held in place by two anchors at intercardinal points off the bow. As illustrated in figure 2-4, these ships can

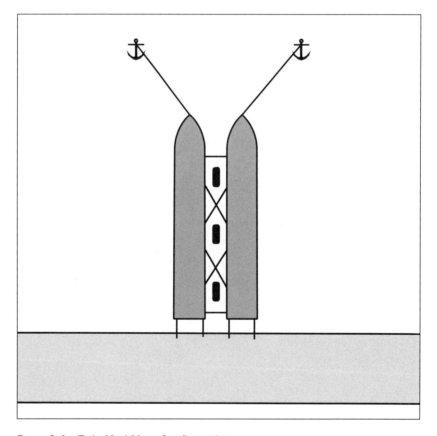

FIGURE **2-4**   Twin Med-Moor Configuration

overcome this limitation by Med-mooring in pairs; two ships can tie up together, one ship providing an anchor off the port bow and the other off the starboard bow. The second limitation is that when anchored in heavy weather the ship will not have a second anchor to place underfoot to dampen bow movement. This limitation will require the ship to resort more quickly to steaming on the anchor, which is described in detail in Chapter Seven.

CONCLUSION

This ship class presents a remarkable challenge for the shiphandler. On the one hand, this lightweight, shallow-draft ship is very susceptible to the uncontrollable forces. Just comparing the large sail area to the small underwater area providing resistance reveals how easily the wind, for example, can carry the ship downwind. On the other hand, the extraordinary thrust-to-weight ratio, maneuverable waterjets, and agile bow thruster produce controllable forces that allow the shiphandler to maintain total control over the ship. Primarily meant to allow the ship to engage in combat in the littorals, this ship's maneuverability and shallow draft also enable it to reach ports unthinkable for larger destroyers and cruisers, particularly in ports with historically unreliable tug services.

This chapter has outlined the uncontrollable and controllable forces that the shiphandler must understand to drive the ship safely and effectively. While the uncontrollable forces are not new, novice and advanced shiphandlers alike are wise to study continually the ship's characteristics and Mother Nature's effects on ship movement. A strong understanding of the controllable forces presented in this chapter will be useful as this discussion focuses on their practical applications—namely, pierwork in Chapter Five, channel driving in Chapter Six, and special evolutions in Chapter Seven.

CHAPTER THREE

Waterjet Vector Management

The most extraordinary shiphandling characteristics of this ship class lie in its maneuverable waterjets. With the ability to direct thrust both forward and aft, as well as left and right, the shiphandler is able to push the stern quickly in any direction. The true capability of the waterjets, however, becomes evident when combining the forces of opposing waterjets to produce truly remarkable outcomes.

As discussed in Chapter One, waterjets allow the shiphandler to employ theoretically infinite combinations to achieve similar ship movements, a fact that makes these waterjets both extremely flexible and potentially hazardous. Creativity has limited value in close quarters; in these situations the conning officer and commanding officer must share clear expectations as to how to use controllable forces to maintain control of the ship. Even though the conning officer may know a combination that would achieve a successful outcome, the shiphandling team as a whole will benefit more from reducing the infinite vector combinations to a set of simple, repeatable methods. The team has the best chance of success when everyone understands the shiphandling plan and the tools employed to overcome challenges. Without this shared understanding, each member of the team must try to read the conning officer's mind, and they lose precious time discerning intentions, time that they could otherwise use to save the shiphandler from a disastrous error.

This chapter will begin with a brief discussion of the pivot point and its importance to shiphandling. Then it will explain how thrust from individual waterjets produces resultant force vectors that cannot be obtained from a single waterjet alone. After presenting a simple method to create repeatable resultant force vectors during close-quarters maneuvering, this chapter will conclude by reviewing basic ship maneuvers and describing how to move the pivot point at will.

PIVOT POINT

Before the use of waterjets to produce resultant force vectors is discussed, the two most important directions of a ship's movement must be explained. *Longitudinal* movement is motion in the same direction of the keel, forward or aft. *Lateral* movement is motion at a right angle to the keel, to port or starboard. For example, longitudinal movement can be created by the ship's propulsion, and lateral movement can be caused by a tug pushing on the port or starboard side.

It is also important to understand how the ship naturally reacts when these forces are applied. Forces, controllable or uncontrollable, that have any lateral component at all usually cause the ship to swivel to one side or the other, and the point about which the ship swivels is called the *pivot point*. When the ship is dead in the water, the natural pivot point is where there is as much underwater surface area forward as aft of the point.[1] Applying a purely lateral force precisely at the pivot point will move the ship laterally without any pivot.

This pivot point will move as a result of three factors: longitudinal force, lateral force, or any force that holds the ship firmly in place. Longitudinal force will move the pivot point in the same direction as the ship's motion. Headway will move the pivot point forward, and sternway will move it aft. Lateral force will move the pivot point in the direction opposite to where the force is applied. For example, ordering a tug to push on the stern will move the pivot point toward the bow, and a tug pushing on the bow will move it toward to stern. Controllable forces holding the ship in place will move the pivot point to the location where the force attaches to the ship. Dropping the anchor will move the pivot point to the hawse pipe, whereas holding a spring line will move it to the ship's bits holding that line.[2] Chapter Five on pierwork will discuss how to use controllable forces to move the pivot point on demand, anywhere along the ship from the bow to the stern, and Chapter Six will discuss pivot-point behavior during channel driving.

Individual and Resultant Force Vectors

Each waterjet exerts a force on the ship by pulling water in, accelerating it with an impeller, and then discharging it from the ship. By Newton's third law, water propelled out of the waterjet results in an equal and opposite reaction by the ship. The waterjet creates a forward force when water is discharged aft of the ship, and it creates an aft force when water is discharged forward underneath the ship. As described in Chapter Two, the direction of water forward and aft is controlled by a reversing plate; at the slowest speeds, some water is directed aft and some is directed forward under the ship. For example, if the reversing plate is positioned such that only 70 percent of the water is discharging aft, 30 percent of the water is being directed forward. In the end, however, a waterjet opened 70 percent produces a forward force, because more water is directed aft than forward.

For this discussion, let's consider force not in its traditional terms, expressed in Newtons or pound-force, but rather in terms of velocity, expressed in speed vectors. We choose this method because it is easier for the novice shiphandler to envision the ship as moving at a certain speed rather than as being acted upon by a certain quantity of force. Through this lens, a waterjet producing a vector representing a five-knot-equivalent force would be one that could eventually accelerate the mass of the ship to a sustained velocity of five knots in the indicated direction.

Consider what happens to force created by a waterjet that is directed to either port or starboard. Instead of directing all the force longitudinally, the waterjet maneuvered to port or starboard will direct some of that force laterally. For example, as illustrated in

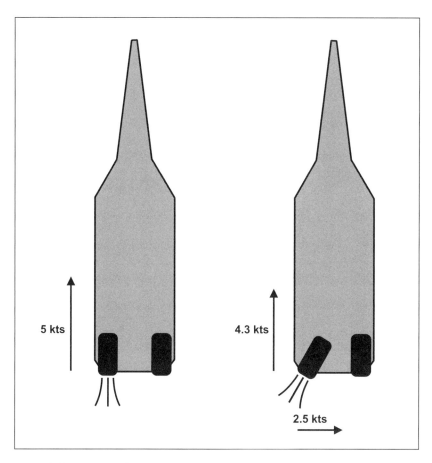

5 kts

4.3 kts

2.5 kts

FIGURE 3-1   Effect of Steering a Waterjet

figure 3-1, if a waterjet that is pushing a five-knot-equivalent force is shifted thirty degrees to port, that five-knot vector will yield a 4.3-knot vector forward and a 2.5-knot vector to starboard, causing the stern to move forward and to starboard.[3]

Combining waterjet vectors reveals the true maneuvering capabilities of this ship class. Consider the above example, but this time in addition to one waterjet thirty degrees to port, another waterjet is directed thirty degrees to starboard with the same five-knot-equivalent force pushing water aft. As illustrated in figure 3-2, both waterjets are still producing a 4.3-knot vector forward, but since the port waterjet is producing a 2.5-knot vector to

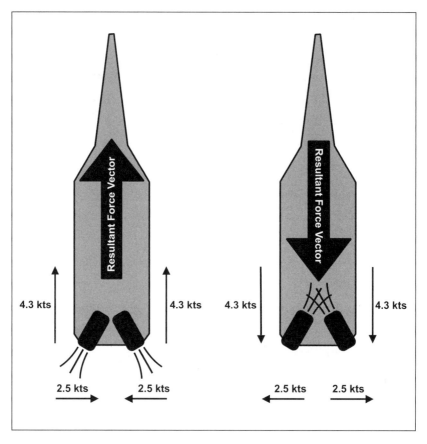

FIGURE 3-2   Combining Force Vectors to Produce a Longitudinal Resultant Force Vector

starboard and the starboard waterjet is producing a 2.5-knot vector to port, the lateral vectors cancel each other out. Adding the 4.3-knot longitudinal vectors together gives the *resultant force vector* from these combined waterjets, an 8.6-knot vector ahead. Now, think about the same two waterjets—one thirty degrees to port and the other thirty degrees to starboard—but with a five-knot-equivalent backing force directing water forward under the ship. Figure 3-2 also illustrates how the lateral vectors still cancel each other out but the two longitudinal vectors added together produce a resultant force vector of 8.6 knots astern.

With the same two waterjets—one thirty degrees to port and the other thirty degrees to starboard—consider what happens when the port waterjet has a five-knot-equivalent ahead force and the starboard waterjet has a five-knot-equivalent backing force. In this case, the longitudinal vectors cancel each other out, but the lateral vectors are additive. Both waterjets create a 2.5-knot vector to starboard, so the resultant force vector pushes the stern at five knots to starboard, as depicted in figure 3-3. Conversely, if the thrust were reversed for each waterjet, so that the port waterjet has a five-knot-equivalent backing force and the starboard waterjet has a five-knot-equivalent ahead force, the resultant force vector would

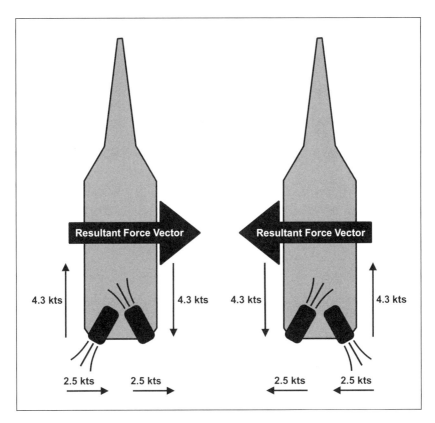

FIGURE 3-3   Combining Force Vectors to Produce a Lateral Resultant Force Vector

be exactly the opposite. The longitudinal vectors would still cancel out, but both waterjets would create a 2.5-knot vector to port, so the resultant force vector would push the stern five knots to port.

While it is not important to memorize the precise values of longitudinal and lateral force vectors for each waterjet angle, table 3-1 further illustrates how changing the waterjet angles can alter the lateral force vector, given constant opposing and canceling longitudinal force vectors. For each of these solutions, the waterjet is producing a five-knot-equivalent thrust.[4]

These examples demonstrate how individual vectors from opposing waterjets combine to produce a resultant force vector on the ship. But we traditionally do not refer to individual waterjets as discharging water at speeds in knots; the above examples are provided only to illustrate the effect of altering waterjets to an angle off centerline. In practice, when giving orders to adjust the waterjets in close-quarters maneuvering, the conning officer orders speeds not in knots but by throttle position. The throttles are numbered from 1 to 10 ahead and 1 to 10 astern in single-digit increments, with 0 indicating "stop;" but since they move smoothly throughout this range of settings, they can be positioned between whole number markings. Standard commands are covered in Chapter Four, but for now, it is sufficient to understand that a throttle in position 1 ahead is referred to as "ahead T1" and position 1 astern as "back T1."

TABLE 3-1 Lateral Force Vectors Resulting from Changing Waterjet Angles, Given Waterjets Producing a Five-Knot-Equivalent Thrust

PORT/STARBOARD OPPOSED WATERJET ANGLES	LATERAL RESULTANT FORCE VECTOR
Five degrees	0.9 knots
Ten degrees	1.7 knots
Fifteen degrees	2.6 knots
Twenty degrees	3.4 knots
Twenty-five degrees	4.2 knots
Thirty degrees	5.0 knots

Toe-In/Toe-Out Method

The last example above reveals an extraordinary capability of the maneuverable waterjets. By simply adjusting the waterjets to *equal and opposing angles* and canceling longitudinal force vectors with *equal and opposing thrust*, the shiphandler can exercise complete control over the lateral resultant force vector on the stern. This can be described as the *toe-in/toe-out method*, as the equal and opposing waterjet angles resemble two feet pivoting on their heels.[5] When a waterjet is positioned outboard of centerline, it is described as *toed-out*; and when it is positioned inboard of centerline, it is *toed-in*. Positioning both waterjets at equal and opposite angles is referred as having them toed-out or toed-in. The exact terminology for each of these positions is described in Chapter Four; table 3-2 is provided for preliminary reference.

One of the best features of the toe-in/toe-out method is that it provides the shiphandler a safety net when moving the stern laterally. With no change of thrust at all, simply reversing the toe angle generates a lateral force vector that acts as a brake, with effects ranging from slowing the stern's movement to halting it completely. Leaving on the counterforce long enough will even reverse the stern's direction. For example, if the waterjets are toed-out fifteen degrees and the stern is moving toward the pier too quickly, centering the waterjets will have the effect of "tapping the brakes," toeing-in the waterjets thirty degrees will amount to "slamming the brakes," and leaving the toed-in configuration on for much longer will accelerate the stern away from the pier. Any

TABLE **3-2** Waterjet Toe-In/Toe-Out Terminology

PORT WATER JET ANGLE	STARBOARD WATER JET ANGLE	TERMINOLOGY
30 degrees left	30 degrees right	Toed-out 30 degrees
15 degrees left	15 degrees right	Toed-out 15 degrees
Centered	Centered	Centered or amidships
15 degrees right	15 degrees left	Toed-in 15 degrees
30 degrees right	30 degrees left	Toed-in 30 degrees

experienced shiphandler on a propeller-driven ship knows the feeling of the stern swinging too quickly toward the pier; it takes quick reactions and a well-trained shiphandling team to shift the rudder and reverse the engines in time to stop the stern's momentum. With this ship class, the conning officer simply needs to reverse the waterjet angle, and the stern will stop smartly.

Mastering control over the rate of lateral movement of the stern is important for handling this ship class. In addition to prevention of a stern allision, which is any shiphandler's concern, precise control over this movement enables the shiphandler to walk the ship laterally. If the stern is moving too quickly, it will outpace the bow, and the ship will twist. The procedure for walking the ship laterally will be discussed in more detail in Chapter Five on pierwork, where the importance of precisely controlling the stern will become more apparent.

When considering the full range of waterjet angles, one important point should be noted regarding the effect of centering the waterjets. Table 3-1 may seem to imply that centering the waterjets would produce a neutral moment—that is, the stern moving neither port nor starboard—but this is not the case. Since the waterjets are offset on opposing sides of the ship's centerline, ordering an ahead engine on one side and a backing engine on the other will produce a slight twisting moment. If the port engine is ahead and the starboard engine is back, the stern will move slightly to port. The opposite configuration, of course, will cause the stern to move slightly to starboard. Therefore, the waterjet angle that produces a neutral moment exists somewhere between zero and five degrees, but from a practical shiphandling perspective finding it is hardly worth the cognitive effort. Even if one successfully calculated the precise angle and the waterjets responded perfectly in settling into that angle, the reality of uncontrollable forces would quickly make the engine order moot. Experienced shiphandlers understand that successful maneuvering is a process of adjusting controllable forces to account for the effects of uncontrollable forces.

In addition, since precision emerges as a key benefit of effectively employing the toe-in/toe-out method, setting the waterjets to cancel the longitudinal force vector is something that must be actively managed by the conning officer. Any experienced shiphandler can attest that ordering engines to the perfect speed is a fool's errand; searching for the perfect thrust setting is a task that consumes considerable attention, for a goal that is rarely attainable. Instead, the practical method of canceling longitudinal force vectors looks more like very slightly jockeying back and forth. If the ship makes no more than 0.1 knots forward and aft, and the aggregate movement throughout the evolution averages zero knots— that is, the ship spends as much time at 0.1 knot ahead as it does at 0.1 knot astern—the ship will have effectively canceled out the longitudinal force vectors.

Consider how the toe-in/toe-out method would look in real time using throttle settings rather than speeds. If the conning officer has the waterjets toed-out fifteen degrees with the port waterjet back T1 and the starboard waterjet ahead T1, the ship may start at zero knots longitudinally but could develop slight headway in time. Altering the backing waterjet to T2 would increase the aft force vector and counter the slight headway. In time, however, the ship might gain slight sternway, in which case the conning officer would simply have to return the backing waterjet to T1. Again, from a practical perspective, as long as the ship averages zero knots over time, the ship has effectively negated the longitudinal vectors.

Some shiphandlers have adopted a method of ordering waterjets by specific reversing-plate percentages. This method would be similar to ordering a specific pitch on a ship with controllable/ reversible-pitch propellers, which is a time-proven procedure. The benefit of ordering specific reversing-plate percentages is that in that way, if it is done quickly and precisely, the conning officer can exercise close control over the ship. The downside, however, is that ordering the helm/lee helm to find a specific reversing-plate percentage with the imprecise throttle lever available can result in throttle hunting—that is, the watchstander moving the throttle by small amounts for an extended period of time trying to find the exact value. This method compromises the conning officer's ability to give an order and observe its effect, since the thrust can be changing by small amounts while the shiphandler is evaluating whether the ordered speed is too much, too little, or just right. Like many other decisions in shiphandling, this choice likely falls under the category of the captain's prerogative, but throttle hunting should be avoided when possible.

This discussion of the toe-in/toe-out method so far has focused on canceling the longitudinal force vectors as a means to walk the ship left and right, but this method can be also used to control the stern's lateral movement while driving the ship ahead or astern. Waterjet vector management dictates that in order to move the ship ahead with two opposing engines, the conning officer simply needs to apply more power on the ahead engine than astern engine. For example, if the waterjets are centered with the port engine back T1 and the starboard engine ahead T1 and the conning officer needs to move the ship forward, ordering the starboard engine to ahead T2 will produce a forward resultant force vector, since the starboard

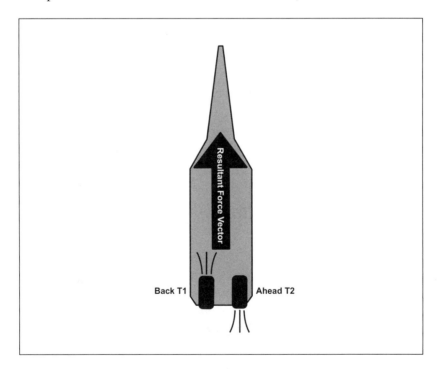

FIGURE 3-4   Driving Ahead against a Backing Waterjet

engine's forward force vector will be greater than the port engine's aft force vector. Increasing the starboard engine to T3 would increase the forward resultant forward force vector even more.

To some shiphandlers, it would seem counterintuitive to drive against an opposing engine. Conventional wisdom dictates that the engines should be ahead when driving ahead and astern when driving astern. Maneuvering the ship in close quarters, however, should motivate the shiphandler to seek maximum control over the stern, and the toe-in/toe-out method offers precise control on demand. In the above example, by keeping the backing engine engaged the shiphandler is able to halt immediately a stern swinging laterally toward danger by simply altering the toe angle. In addition, if the ship is moving too quickly ahead in a slip, instead of reducing the ahead engines and waiting for the water's resistance to slow the ship the conning officer can increase the backing engine to apply braking power to bring the ship smartly to a safe speed. The value of this precise control over the ship by using the toe-in/toe-out method cannot be overstated.

While the toe-in/toe-out method is reliable and proven, it is not the only approach that can be trusted in close quarters. The strengths of this method have been detailed already, but the flexible nature of waterjet vector management makes it appropriate to discuss at least one alternative. The theoretical math behind waterjet vector management makes it possible to use an infinite number of waterjet combinations to achieve the same resultant force vectors. As in the toe-in/toe-out method, the goal is to cancel out undesirable force vectors and accentuate desirable force vectors to produce the required controllable force.

Consider one alternate configuration, one in which the shiphandler divides the force vector duties between two waterjets, using one waterjet to create the lateral force vector and the other waterjet to cancel the undesired longitudinal force vector. For example, if trying to walk the ship to starboard with a port waterjet pushing a five-knot-equivalent force, toeing-out the port waterjet thirty degrees will produce a 4.3-knot vector forward and a 2.5-knot vector to starboard. The conning officer then leaves the starboard waterjet centered with 4.3-knot-equivalent backing engine to cancel the forward force vector. The resultant force vector is 2.5 knots to starboard. Since the waterjet cannot be toed-out more than thirty degrees, the only way to increase the lateral force vector would be to increase the discharge force from the waterjet. Increasing the thrust to a ten-knot equivalent would produce a five-knot vector to starboard, but it would also generate an 8.6-knot forward vector. In order to prevent the ship from surging forward, the conning officer can increase the starboard backing waterjet to an 8.6-knots equivalent to cancel the longitudinal force vector, as illustrated in figure 3-5.

Consider the opposite situation, in which the conning officer needs to *reduce* the stern's lateral speed. Instead of decreasing the engine power from its five-knot-equivalent force, the shiphandler

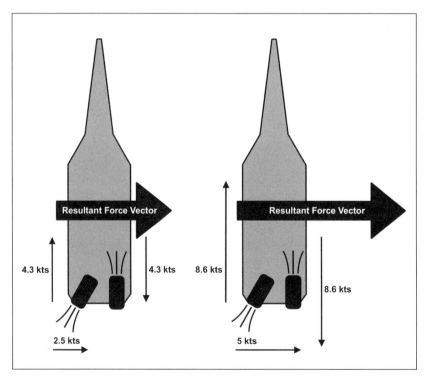

FIGURE 3-5   Producing a Lateral Resultant Force Vector by Steering Only One Waterjet

can simply reduce the toe angle. Changing the waterjet toed-out angle from thirty degrees to fifteen degrees will reduce the lateral force vector to 1.3 knots, but since the waterjet is aligned closer to the centerline, the forward force vector will increase concurrently. In order to maintain zero longitudinal movement, the conning officer must increase the thrust of the starboard backing engine.

The theoretical math behind waterjet vector management reveals, as noted, an infinite number of combinations that can achieve the same resultant force vectors. The toe-in/toe-out method is presented here as the most reliable option, because it is *simple, repeatable,* and *proven.* As discussed in the preface, however, the captain's prerogative prevails, and his or her comfort level with a particular approach must be given the most serious regard. As we have seen, the most important principle in safely handling the ship is that there must be clear expectations between the conning officer and the captain about what shiphandling orders will be applied during a given evolution. Unless he or she knows what orders to expect, the captain cannot serve as an effective safety net for a less experienced conning officer.

THE LIMITS OF MATH IN PRACTICAL SHIPHANDLING

Some argue that all good shiphandlers must begin by becoming very good at the mathematics behind shiphandling forces, a view that serves as a barrier against junior shiphandlers who have uncomfortable relationships with trigonometry. Whether one relates better to the art or to the science of shiphandling, the first step toward becoming an advanced shiphandler is to find a relationship with the ship's movement that makes sense. Shiphandlers who are uncomfortable with math in general cannot rely on their ability to calculate sines and cosines under pressure. They must find a way that works for them, practice it to solidify the concepts, and then rely on it when driving the ship.

For example, when discussing these concepts with junior shiphandlers, it is apparent that some officers particularly uncomfortable with vector math have found success thinking of waterjet management in terms simply of *pushing* and *pulling* water through the waterjets. Even though the path of water through the intake does not actually change whether the engine is ahead or astern—the reversing plate simply changes the direction of discharge—the concept is reliable. Ordering an ahead engine is like pushing water out the back of the ship and thereby pushing the ship ahead. Conversely, ordering a backing engine is like pulling water from behind the waterjet and so pulling the ship astern. As the waterjet angle increases, the pushing and pulling forces pivot accordingly.

In fact, there are limits to the value of math in practical shiphandling. Throughout the course of a docking evolution, for example, the dynamic nature of the uncontrollable forces produces varying ship movements, and the shiphandler must dynamically counter them with controllable forces. No precise math solution will survive first contact with the uncontrollable forces; the conning officer must frequently adjust the controllable forces to maintain control over the ship.[6]

BASIC SHIP MANEUVERS

As described above, the flexible nature of waterjet vector math allows the shiphandler to employ countless waterjet combinations to achieve the same results, but conning officers should have tool kits of basic shiphandling maneuvers at their disposal. These maneuvers serve as starting points for novice shiphandlers, but they also serve advanced shiphandlers well when hazardous circumstances shorten the timeline for creative thinking. The most basic waterjet combination is often the best prescription when reaction time is short. To be clear, many other combinations may suffice as well, but these will serve as a "go to" list.

Twisting

The twist is one of the most basic maneuvers in shiphandling, and the ability to produce immediate turning thrust with the waterjets while dead in the water allows this ship class to twist on short order. In this maneuver, the bow and stern move in opposing directions as the ship's pivot point moves forward. In a port twist, the bow moves to port, and the stern moves to starboard. Conversely, a starboard twist moves the bow to starboard and the stern to port. Consider two ways in which this ship class can twist.

Toe-In Twist

To twist the ship to port, toe-in both engines, with the starboard engine ahead and the port engine back, as illustrated in figure 3-6. A starboard twist requires the opposite—the port engine ahead and the starboard engine back. The rate of twisting can be controlled with the waterjet angle. For the maximum rate of turn, toe-in the engines thirty degrees; to slow the rate of turn, simply decrease the toe angle. With the waterjets toed-in thirty degrees, the shiphandler can further increase the rate of turn by increasing thrust on each engine. In order to keep the ship stationary, the thrust should be adjusted such that the longitudinal force vectors cancel out.

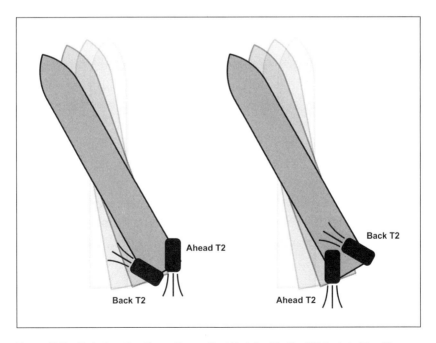

FIGURE 3-6   Twisting Configurations: Port Twist with the Waterjets Toed-In
(Left) and Toed-Out (Right)

Toe-Out Twist

The ship can also be twisted to port with both waterjets toed-out, port engine ahead and starboard engine back, which is also depicted in figure 3-6. Conversely, to twist the ship to starboard, the shiphandler would order the port engine back and the starboard engine ahead. As in the toed-in twist, the rate of turn can be controlled by waterjet angle and, if necessary, engine thrust. Increasing thrust on each engine with the waterjets toed-out thirty degrees will further increase the rate of turn. Again, the throttle settings should be adjusted to ensure that the longitudinal force vectors cancel out in order to keep the ship stationary.

Laterally Walking

"Walking the ship" is moving it to port or starboard with zero longitudinal movement, with the bow and stern moving at the same lateral speed. This lateral movement is a distinctive feature of waterjet shiphandling and provides maneuverability that this ship class needs to operate in austere ports, where tug services may be more a hazard than help. In these situations, the ship can settle adjacent and parallel to the pier and then walk slowly toward it until close enough to put lines over. The conning officer can then use the walking configuration to hold the ship against the pier until all lines are secured. Since the goal during a walk is to match the lateral speeds of the bow and the stern, the best measure of success is the ship's heading. A steady heading while closing or opening the pier is one of the indicators that the ship is walking laterally.

The *Independence* variant has the benefit of a bow thruster, which offers precise control over movement on the bow and an ability to build lateral speed quickly. Without the thruster, building lateral speed is not as easy, particularly if uncontrollable forces are working against the direction of the intended walk, but it is possible under the right conditions. To understand how the ship can walk laterally without a controllable force on the bow, consider the earlier discussion regarding the twisting moment caused by centered waterjets. The waterjets are offset on opposing sides of the ship's centerline, so ordering an ahead engine on one side and a backing engine on the other will produce a slight twisting moment. If the port engine is ahead and the starboard engine is back, the stern will move slightly to port and the bow will pivot to starboard. Toeing-out ten degrees, however, will reverse this lateral force: the stern will move to starboard, and the bow will pivot to port. At some angle between centered and toed-out ten degrees, the port force vector still moves the stern to starboard but the force is not great enough to cause the bow to pivot about the pivot point. At this angle, both the bow and stern are moving laterally at the same speed.

Notionally, this angle is six or seven degrees, but caution is due on this point; novice shiphandlers should not believe that these mathematically calculable angles are consistently achievable in practice. Even if one successfully found the precise angle, the helm/lee helm adjusted the waterjets precisely, and the waterjets settled at the angle precisely, the hope of mathematical shiphandling would fade as uncontrollable forces weighed in. Experienced shiphandlers understand that successful maneuvering is a process of adjusting controllable forces to account for the effects of the uncontrollable forces. In practice, walking the ship is an exercise in controlling the bow and stern movement so that the *average* heading remains the same. If the ship's heading begins at 180 degrees true and varies between 178 and 182 but averages 180 throughout the maneuver, the ship is considered to be walking laterally.

Since walking the ship laterally calls for zero longitudinal movement, the shiphandler must watch the ship's speed carefully to ensure that it does not surge in either direction. Even speeds as low as a tenth of a knot add up in distance over time, which could result in the ship being out of position alongside the pier or even standing into danger. Again, longitudinal movement can be controlled by increasing forward or aft thrust to produce longitudinal vectors that cancel out. For example, if the conning officer begins walking with the port engine ahead T1 and the starboard engine back T1and the ship begins to move ahead at 0.1 knots, the conning officer can order the starboard engine back T2 to check the forward motion. If sternway begins to develop, the starboard engine should be returned to back T1. Should this sternway persist, the conning officer can then order the port engine ahead T2 to check the aft motion. As long as the ship's longitudinal speed averages zero knots throughout the maneuver, the ship is considered to be walking laterally. Below are two ways to walk the ship.

Toe-Out Walk

To walk the ship to port, both engines are toed-out six degrees, with the starboard engine ahead and the port engine back. If the heading increases, the stern is outpacing the bow, so the conning officer must reduce the toe angle. If the heading decreases, the ship has begun twisting to port, so the conning officer must increase the toe angle. Reversing the engine thrust will walk the ship to starboard, with the port engine ahead and the starboard engine back. The practical application of this maneuver will see the shiphandler switching between toed-out ten degrees, toed-out six degrees, and waterjets centered in order to achieve an average desired heading.

To walk the ship to port with a bow thruster, toe-out both engines with the starboard engine ahead, the port engine back, and the bow thruster pointed to 270 degrees relative. Walking the ship to starboard would require reversed engine thrust—the port engine ahead, the starboard engine back, and the bow thruster pointed to 090 degrees relative, as illustrated in figure 3-7. The shiphandler controls the lateral speed of the stern by increasing or decreasing the toe angle and that of the bow by increasing or decreasing power with the bow thruster.

Toe-In Walk

Walking laterally to port in a toed-in configuration requires the port engine ahead and the starboard engine back. Begin with the waterjets toed-in six degrees; if the heading increases, the stern is moving more quickly than the bow, and the toe angle must be reduced. If the heading decreases, the ship has transitioned to a port twist, and the waterjets must be toed-in more. The main difference between the toe-in and toe-out walk is that the twisting moment is reversed when the waterjets are centered. For example, whereas centered waterjets in a toe-out walk to port cause the ship to twist to port, centering the waterjets during a toe-in walk to port will cause it to twist to starboard. Therefore, to halt momentum when the stern is outpacing the bow during a toe-in walk to port, the waterjets may need to be toed-out slightly. Still, the shiphandling approach is the same for a toe-in walk: if the bow is outpacing the stern, toe-in more; if the stern is outpacing the bow, toe-in less, even to the point of toeing-out if necessary.

To walk the ship to port with a bow thruster, the waterjets should be toed-in with the port engine ahead, the starboard engine back, and the thruster pointed to 270 degrees relative. Walking the ship to starboard with the waterjets toed-in would simply require reversing the thrust direction, with the starboard engine ahead, the port engine back, and the thruster pointed to 090 degrees relative, as depicted in figure 3-7. Again, the shiphandler controls

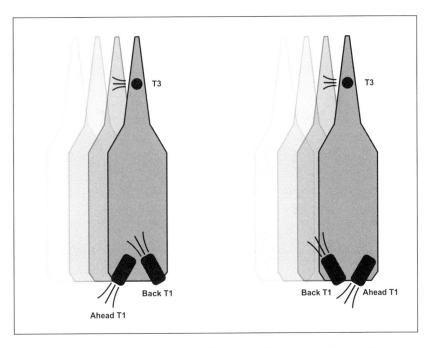

FIGURE 3-7   Walking Configurations: Walking to Starboard with the Waterjets Toed-Out (Left) and Toed-In (Right)

the lateral stern speed with the toe angle and the lateral bow speed with the bow thruster.

Crabbing

"Crabbing" is a common ship maneuver that combines walking laterally and moving either ahead or astern. In other words, crabbing is moving diagonally without altering the ship's heading. While this ship class is able to walk laterally without longitudinal movement, maneuvering at right angles is not always efficient in practice. The crabbing maneuver is particularly useful during docking and undocking to expedite the evolution at a controlled speed, and it can be accomplished with or without a thruster.

In its most basic form, crabbing the ship begins by laterally walking the ship as described above. As the ship begins moving laterally, the conning officer increases thrust on one of the engines to produce a resultant force vector in the intended direction. To crab the ship to starboard with headway, the waterjets can be toed-out with the port engine ahead and starboard engine back. Forward motion is developed by pushing harder on the ahead engine than on the astern engine. For example, increasing the port ahead engine to T2 while maintaining the starboard backing engine at T1 will build slight headway. Conversely, crabbing the ship to port with sternway in a toed-out configuration can be accomplished with the port engine back T2 and the starboard engine ahead T1. Longitudinal speed is carefully controlled by adjusting the ahead or backing engines to achieve the required resultant force vector, and the ship's heading is maintained by adjusting the toe angle, as described above for a lateral walk.

The shiphandler can crab the ship toward any of the four intercardinal points with even more precision by employing a bow thruster. Pointing the thruster in the direction of the intended lateral movement will help the conning officer maintain the ship's heading throughout the maneuver, as illustrated in figure 3-8. Additionally, the ship can be crabbed using a toed-in configuration simply by reversing the thrust direction.

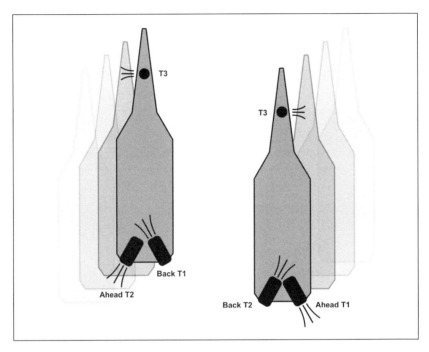

FIGURE 3-8   Crabbing Configurations: Forward-Starboard (Left) and
Aft-Starboard (Right), Both with a Toed-Out Configuration
and Using a Bow Thruster

THE SIX SHIPHANDLING ACTIONS

For the novice shiphandler, waterjet vector management may appear enormously challenging, and it could seem impossible to attempt these mental gymnastics while maneuvering a warship just a few feet from the pier. All shiphandlers, from novice to advanced, should think carefully through their docking or undocking plan long before the evolution begins. Once the shiphandling has begun, the conning officer really needs to remember only the *six ship-handling actions*, which are derived from the following six questions:

1. What makes the stern go to port?
2. What makes the stern go to starboard?
3. What makes the bow go to port?
4. What makes the bow go to starboard?

5. What makes the ship move ahead?

6. What makes the ship move astern?

Keeping in mind the answers to these six simple questions, the shiphandler does not need to rethink the vector math with each waterjet or engine change. Shiphandling is merely taking action, then observing the reaction, then taking another action, then observing the reaction—over and over. A conning officer who writes down the six actions above will have every shiphandling order in hand before ever taking the watch.

Consider these questions with a ship mooring to a pier along the starboard side, using the toe-in/toe-out method to walk the ship laterally to starboard. In this example, the shiphandler will have a thruster (or tug for the *Freedom* variant) available for maximum control over the ship. Walking the ship to starboard with a toed-out configuration would require the port engine ahead, the starboard engine back, and the thruster trained to 090 degrees relative (or a tug pushing ahead to starboard). For this docking, table 3-3 on the six shiphandling actions would be useful for quick reference:

When beginning the docking evolution with the ship stopped parallel to the pier, the conning officer engages the waterjets, with

TABLE 3-3 Six Shiphandling Actions for Walking the Ship Laterally to Starboard with the Port Engine Ahead, Starboard Engine Back, and Employing a Thruster (or Tug)	
SIX SHIPHANDLING QUESTIONS	**SIX SHIPHANDLING ACTIONS**
1. What makes the stern go to starboard?	Toeing-out waterjets.
2. What makes the stern go to port?	Toeing-in waterjets.
3. What makes the bow go to starboard?	Thruster (or tug power) to 090 deg. relative.
4. What makes the bow go to port?	Thruster (or tug power) to 270 deg. relative.
5. What makes the ship move ahead?	Increasing the port ahead engine.
6. What makes the ship move astern?	Increasing the starboard backing engine.

the port engine ahead T1, the starboard engine back T1, and the thruster ahead T3 (or tug pushing to starboard). Now that action has been ordered, the shiphandler observes the reaction. If the stern is not moving to starboard as quickly as the bow, the waterjets should be toed-out more. If the stern is approaching the pier dangerously fast and must be pulled away, the waterjets should be toed-in to make the stern move back to port. Similarly, if the bow is not moving to starboard as quickly as the stern, thruster (or tug) power should be increased toward 090 degrees relative; if it is moving too quickly toward the pier and must be pulled back, the thruster should be trained to 270 degrees relative (or tug ordered away to port). The shiphandler also watches the longitudinal movement of the ship to ensure that the ship is moving neither ahead nor astern. If the ship begins surging ahead, the port engine should be increased to back T2; if the ship is falling astern, the starboard engine should be increased to ahead T2.

This sequence of orders might seem self-evident to even the most inexperienced shiphandler. However, relying on common sense or memory in the first or second year of shiphandling is a risky game. Thinking through the six shiphandling actions ahead of time—even arriving on the bridge holding a "cheat sheet" with these six answers—will show that the conning officer is serious about the business of shiphandling. It might even save the ship from an allision should the undocking go awry.

One key point to take away from the list of the six shiphandling actions is that it separates movement of the stern from that of the bow. To be sure, this does not mean that forces applied to the stern have no effect on the bow or vice versa. Recalling the earlier discussion on the pivot point, we know that the ship will pivot in such a way that forces applied to once side of the pivot point will cause a proportional and opposite movement on the other.[7] So, how is the shiphandler to resolve the clear relationship between the bow and stern, based on the pivot point, when the listed six shiphandling actions separate bow and stern movement?

One way to simplify the relationship is to consider the bow and stern forces as interrelated but not interchangeable. To understand the interrelationship, the shiphandler should anticipate that moving the bow to port with the thruster or tug will cause the stern to move to starboard. Conversely, moving the stern to starboard with the waterjets will cause the bow to move to port. The shiphandler, however, should not use the bow thruster to move the stern or use the waterjets to move the bow but should simply expect that one will affect the other.

A very straightforward way to achieve such simplicity in managing the controllable forces is to divide the ship in half, designating the bow thruster or tug to control the bow and the waterjets to control the stern. With this approach, the shiphandler can easily and predictably react to observed movements. If the bow is not moving in the right direction, simply train the bow thruster (or tug) in the desired direction. Similarly, if the stern is not moving properly, toe-out or toe-in the waterjets to move it in the correct direction. This approach is a slight departure from using the waterjets alone to walk the ship laterally, but for the novice shiphandler a little simplicity goes a long way.

If we know that applying a lateral force offset from the pivot point will cause a certain movement on the other end of the ship, we might ask why that force cannot be used to achieve the desired effect. Why not pull the bow to port in order to move the stern to starboard? The difference is small but important. To understand why this is the wrong tool, consider the effect of applying a lateral force on the ship, in particular its effect on the pivot point. Applying a lateral force offset from the ship's pivot point will cause the pivot point to shift away from the position where the force is applied. So, using a bow thruster or tug on the bow will shift the pivot point aft. The stronger the force, the farther astern it will move. Let's say that pushing on the bow has moved the pivot point to a location three-quarters of the ship's length from the bow. With the pivot point in this position during a lateral walk toward a pier, the conning officer would have to move the bow three feet to achieve one foot of stern movement. If the conning officer needed to move the stern twenty feet closer to the pier, the bow would have to be moved sixty feet away from the pier. This approach may achieve the desired effect, but it is a very inefficient way to handle the ship.

To exercise total control over the ship, the shiphandler must be able to distinguish between *expecting* the thruster to move the stern and *using* the thruster to move the stern. The former is indicative of an advanced shiphandler who anticipates the consequences of controllable forces on the ship; the savviest of them take preemptive action to check the consequences. The latter is indicative of a novice shiphandler who concludes that the cause-and-effect relationship makes for sensible shiphandling. Simply put, employing bow forces to move the stern or stern forces to move the bow is using the wrong tool for the job.

There are two exceptions to this rule. The first exception, as stated above, is laterally walking the ship with the waterjets alone. In this case, the waterjets are moving the bow and stern together. The second exception is twisting, where the shiphandler is in fact seeking to move the bow and stern inversely. Since the twist is accomplished by applying a lateral force either forward or aft of the pivot point, the ship can be twisted with either the waterjets on the stern, the thruster or tug on the bow, or both in tandem.

MOVING THE PIVOT POINT

This chapter began by defining the pivot point and conceptually discussing the longitudinal, lateral, and stationary forces that cause the pivot point to move. Now, with the benefit of understanding how controllable forces can be applied to this ship class, consider how these forces move the pivot point in the course of normal shiphandling maneuvers.

When the ship is stationary, recall, the pivot point marks the location along the keel with as much underwater surface area forward as aft.[8] The pivot point shifts as a result of longitudinal movement, lateral application of force, and application of any controllable force that holds the ship firmly in place. Longitudinal movement shifts the pivot point in the same direction as the ship's motion; headway moves the pivot point forward, and sternway moves it aft.

The lateral application of force will move the pivot point in the direction opposite to the point where the force is applied; ordering a tug to push on the stern will move the pivot point toward the bow, and a tug pushing on the bow will move it toward the stern. Finally, controllable forces holding the ship in place will move the pivot point to the position where the force, such as a mooring line holding the bow to a pier, attaches to the ship.[9]

To appreciate the effects of ship's motion on the pivot point, consider how the pivot point will behave during undocking when proceeding out of a slip. In this example, the ship is moored with its starboard side toward the pier and the bow pointed toward the slip's exit. Laterally walking the ship to port, with equal force on the bow and stern and zero longitudinal motion, will not appreciably affect the pivot point, which should remain in its natural position. Once the ship is far enough from the pier and the conning officer proceeds ahead, this forward movement will cause the pivot point to shift forward along the ship's axis, so uncontrollable forces will affect the stern more than the bow. The faster the ship travels ahead, the farther forward the pivot point moves.

This movement is particularly important to keep in mind when turning into a channel that runs perpendicular to the pier, because putting over the waterjets applies a lateral force aft that will shift the pivot point even farther forward. If the ship is making significant way, or if the conning officer uses considerable lateral force astern, the lateral movement of the stern will be much more than that of the bow. For example, if the pivot point has shifted to one-quarter the ship's length from the bow, the stern will laterally move three feet for every foot the bow moves. Most novice shiphandlers will spend more time looking ahead than astern when driving into the channel; more experienced shiphandlers know to watch the stern swing to ensure that the ship remains clear of adjacent piers and buoys.

Conversely, think through pivot-point behavior when backing out of the slip. A ship moored starboard side toward the pier with its stern toward the slip entrance would have to back out into the channel before proceeding out of port. Again, laterally walking the ship to port, with equal force on the bow and stern and no longitudinal motion, does not alter the pivot point from its natural position. After moving toward the center of the slip, the conning officer will back toward the channel, which will cause the pivot point to shift aft along the ship's axis, so uncontrollable forces will affect the bow more than the stern. As in moving ahead, the distance the pivot point shifts is proportional to the ship's speed. Once the turn commences, however, putting over the waterjets applies a lateral force on the stern that will shift the pivot point forward. This combination of factors—astern movement that shifts the pivot point aft and the lateral force on the stern that shifts the pivot point forward—will have a canceling effect, whereby the pivot point will sit closer to its natural position. In this case, the difference between stern swing and bow swing will be much less significant during the turn.

Discussions about the ship's pivot point often consider its movement as a consequence rather than a shiphandling objective. The extraordinary maneuverability of the *Independence* variant has turned that perspective on its head, thanks primarily to the capability of the bow thruster in combination with the waterjets. This ship is so maneuverable that the shiphandler can *choose* where to place the pivot point using the ship's controllable forces alone.[10] The true value of this capability becomes apparent during docking and undocking, when the shiphandler is constrained from producing significant longitudinal motion. To appreciate the ship's ability to move the pivot point along its entire length, consider a simple shiphandling evolution in which the ship must alter its heading 180 degrees. In each of the three examples below the pivot point is in a different position but still facilitates bringing the ship about.

Pivot Point Amidships

Much as in a traditional twist, consider this problem like inserting a pushpin through the center of the ship and into the harbor floor. To achieve this pivot point position, the bow must be moved at the same rate as the stern but in the opposite direction. This maneuver can be achieved, for example, by toeing-out both waterjets, with the starboard engine ahead and the port engine back at equal thrust. To keep the pivot point from sliding forward—away from the applied lateral force from the waterjets—the thruster should be trained to 090 degrees relative, with just enough power to match the lateral movement of the stern. Keep in mind that pointing the lateral thrust vectors in opposing directions on opposite sides of the pivot point will have an additive effect on the rate of turn. Applying a force on the stern that causes a rate of turn of fifteen-degrees per minute and a force on the bow that also produces a rate of turn of fifteen-degrees per minute will result in a thirty-degree rate of turn. Accordingly, only half the power on each end of the ship will be required to achieve the same rate of turn.

Pivot Point on the Bow

Consider the same maneuver—bringing the ship about 180 degrees—but moving the pivot point to the bow. This time, imagine inserting a pushpin through the forecastle and into the harbor floor. The stern will bear the burden of movement, as it swings through a wide arc to point the ship in the right direction. This force will shift the pivot point forward, but only to a position where the bow moves in the opposite direction more slowly than the stern. Therefore, a lateral counterforce must be applied to hold the bow in place. To effect this pivot-point location, the conning officer can toe-out both waterjets, with the starboard engine ahead and the port engine back at equal thrust. To check the bow motion, the shiphandler points the thruster to 270 degrees relative and applies just enough thrust to cancel out the bow's starboard movement. Holding the bow to negligible lateral movement and the ship to negligible longitudinal movement will shift the pivot point all the way to the bow. This maneuver is useful in circumstances where the ship must come about but bow movement is constrained.

Pivot Point on the Stern

Now think through the same scenario, but this time move the pivot point to the stern—as if a pushpin is inserted through the flight deck. In this case, the bow will carry out the movement as it swings through a wide arc toward the intended track. To move the pivot point to the stern, a great deal of lateral force on the bow is required. This force will shift the pivot point aft, but only to a position where the stern moves in the opposite direction more slowly than the bow; a lateral counterforce must be applied to hold the stern in place. In this example, the conning officer could train the thruster to 270 degrees relative at maximum power, with both waterjets toed-out with the starboard engine ahead and the port engine back at equal thrust. The toe angle determines the amount of lateral force applied to the stern, so the conning officer must adjust the toe angle to provide just enough lateral force to port to cancel the pivot induced by the bow. Placing the pivot point on the stern is useful for situations when the ship needs to come about but stern movement is constrained.

As previously mentioned, calculating mathematical solutions for these offsetting forces is of limited value, as the shiphandler is creating these effects in conjunction with ship movement produced by uncontrollable forces. In practical application, the conning officer simply needs to be aware of the effect of controllable forces on the pivot point and to apply them deliberately to achieve a desired effect. The shiphandler's transition from novice to expert is marked, among other qualities, by a steady decrease in surprises. Once the shiphandler understands the relationship between ship movements and pivot-point shifts, controllable forces can be applied preemptively to move the pivot point precisely when desired.

CONCLUSION

Building on Chapter One's introduction to this ship class and Chapter Two's overview of the controllable and uncontrollable forces, this chapter introduced the shiphandler to employing waterjets to maximize the ship's maneuverability. Effective shiphandling with waterjets begins with a firm understanding of waterjet vector management. Once the shiphandler understands how to use the waterjets to produce resultant force vectors, maneuvers previously considered impossible suddenly become routine.

The nearly infinite variety of waterjet combinations, however, is both the greatest strength and most severe liability of this ship class. Allowing the novice shiphandler to roam about these combinations in search of unorthodox solutions prevents the more experienced shiphandlers—namely, the commanding officers—from serving as effective safety nets. Practicing *simple, repeatable methods* gives the novice shiphandler reliable tools for the pressure-filled environment of close-quarters maneuvering, and it builds within the shiphandling team a shared understanding that allows the more experienced officers to catch errant commands before they produce an allision. The toe-in/toe-out method is just one example of how the shiphandler can tame the countless ways to perform basic maneuvers, such as twisting and laterally walking the ship. Similarly, the six shiphandling actions, when considered ahead of time and written down for handy reference, can eliminate errors induced by the pressure of the moment. The ability to grasp intuitively how to position the waterjets will develop over time, after which these tools will have decreasing value and it will become second nature to move the pivot point at will using the ship's controllable forces.

CHAPTER FOUR

Standard Commands

S ome aspects of shiphandling are universal, regardless of the ship class, and the use of standard commands to facilitate safe control of the ship is certainly among them. Standard commands ensure that the conning officer's intentions are clearly understood by the helm/lee helm, but the broader value in using standard commands benefits the other watchstanders on the bridge. Standard commands institutionalize a common language for controlling the ship, so anyone within earshot of the conning officer knows his or her shiphandling intentions. All watchstanders on the bridge develop improved situational awareness, understanding what is required from the ship and anticipating how the ship should react.

This collective understanding permits each person to contribute to the shiphandling task. For example, if the navigator sees that there is very little room on the port quarter between the ship and a buoy and hears the conning officer give a standard command that will swing the stern to port, the navigator can warn that the ship is in danger. If the engineering officer of the watch hears the conning officer order the starboard diesel back T2 but sees the waterjet open only to back T1, the engineer can alert the shiphandling team that either the helm/lee helm has positioned the throttle incorrectly or an engineering casualty is affecting the waterjet. This situational awareness among the watchstanding team, facilitated by standard commands, is priceless.

Finally, standard commands are a critical factor allowing the commanding officer to oversee shiphandling in close quarters. The

U.S. Navy has a long and important tradition of teaching ship-handling to junior officers by means of hands-on training. Although we now benefit from realistic simulators, no computer simulation can replicate the pressure of close-quarters situations like docking, undocking, and underway replenishment. Novice shiphandlers in pressure situations can be expected to confuse engine direction and waterjet angles; the commanding officer is the safety net that prevents an honest mistake by a young officer from damaging the ship, an oiler, or an adjacent pier. If the commanding officer is to provide this backup effectively, engine and waterjet orders must be understood before they are enacted, and standard commands make this possible. Requiring the conning officer to issue these orders out loud to a helm and lee helm, and in simple, commonly under-stood language, will permit the commanding officer to intercede should an error occur.

To experienced shiphandlers on propeller-driven ships, these principles are probably clear, even self-evident; the Navy has embraced them throughout the fleet. This perspective, however, has been challenged by the emergence of a maneuverable ship with controls designed to be physically handled by the conning officer. The ship's console, based on a commercial method of driving ships, is designed with the shiphandler seated within arm's reach of the controls. This approach enables manning reductions during open-ocean transits, where the conning officer does not require sepa-rate helm and lee-helm watchstanders to change the ship's course and speed. In such situations, where erroneous engine orders pose less risk, the conning officer can simply reach over and make the changes.

This risk is increased substantially, however, by the reduced reaction time of close-quarters maneuvering. Consider the predic-ament of the commanding officer should the conning officer make changes to the waterjet configuration without verbalizing them. The commanding officer must first observe that the ship is mov-ing in the wrong direction, then either query the conning officer about the engine configuration or look at the throttles, and only then, having determined the engine and waterjet configuration, compare the conning officer's orders to the prevailing conditions. This is precious time lost in circumstances when reaction time is already reduced.

Novice shiphandlers may think that having to use standard commands slows down the shiphandling process if they could themselves simply adjust the waterjets, but the advanced shiphandler knows that effective control is more than just expediency. The situational awareness throughout the shiphandling team facilitated by standard commands and the safety net provided by the commanding officer are invaluable for safe maneuvering in close quarters.

BASIC PRINCIPLES OF STANDARD COMMANDS

While the terms used for standard commands on this ship class differ from propeller-driven ships, certain principles still apply. The following three principles of standard commands are worth reviewing, particularly when learning to control the ship verbally for the first time.

Clarity Is Critical

When orders are issued to control the ship in close quarters, the helm/lee helm must clearly understand them. There is no room for ambiguity when applying controllable forces, so the conning officer must state orders clearly and in a firm, confident tone. An uncertain tone can have the unintended consequence of delaying execution if the helm/lee helm hesitates to apply it. The helm/lee helm may wait to hear the order again or give the conning officer time to retract a presumably erroneous order, or worse, wait to see if the commanding officer overrides it. In any situation, the conning officer's order must inspire confidence in the watchstander so that it will be executed without delay.

Standard Commands Must Be Standard

Someone not previously exposed to standard commands may consider their structured nature arbitrary and unnecessary. If the meaning is conveyed, why should the order of the words matter? In reality, it is a matter of expectations. If the helm/lee helm is expecting maneuvering instructions to come in a certain sequence, but the conning officer delivers them in a different order, the helm/ lee helm will likely pause for a moment, internally translating the order from what was heard to what should have been heard. These interruptions degrade reaction time and increase the probability of an order being misunderstood. We can get orders translated into action quicker and with fewer errors by following an agreed and simple pattern of communicating—the time-tested tradition of order, repeat-back, confirmation, and acknowledgment. The sequence would resemble the example below:

> Conning officer: "Toe-out diesels fifteen degrees."
>
> Helm/lee helm: "Toe-out diesels fifteen degrees, aye."
>
> Helm/lee helm: "Diesels toed-out fifteen degrees."
>
> Conning officer: "Very well."

The structured nature of standard commands does not allow for much creativity in word choice. Similarly, the helm/lee helm should not be required to distinguish between words with similar meanings. Engine directions like "back" and "astern" may mean the same thing, but if the helm/lee helm has been trained to expect the word "back," the conning officer should use it. In addition, standardization drives the shiphandling team toward brevity, since good standard commands eliminate unnecessary words. Brevity is important in close-quarters maneuvering, because minimizing words improves reaction time and reduces the probability of errors.

If in Doubt, Seek Confirmation!

Close-quarters maneuvering inevitably changes the watchstand-ing environment on the bridge. The decreased reaction time and the increased consequences of error will have an effect that can allow confusion to creep in, so every watchstander must pay particular attention to ensure that orders are clearly understood.

A helm/lee helm not sure of the order given must ask that it be repeated, employing the commonly used phrase, "Orders to the helm?" Conversely, after a sequence of numerous engine and waterjet angle orders, the conning officer may forget the last order given. Forgetting engine and waterjet positions is something to be avoided, but the greater offense is pretending to know. In this case, the conning officer should ask, "How are my engines?" The helm/lee helm would then give a succinct summary, such as, "Waterjets are toed-out fifteen degrees, port ahead one, starboard back one."

To be sure, the conning officer should not use this query when in doubt that the helm/lee helm has positioned the throttles as directed, because it could result in an unnecessary exchange. For example, if the conning officer has ordered both engines back T2 but suspects that they are only back T1, the shiphandler may ask, "How are my engines?," which will prompt the helm/lee helm to report that they are back T1. In this case, the conning officer still has to reissue the order for both engines back T2; so what was the good of asking the status of the engines? If the conning officer knows what should have been done with the throttles, simply reissuing the order will promptly rectify the situation. If the engines are in fact in that position, the helm/lee helm simply responds, "Both engines are *already* back T2."

STANDARD COMMANDS

In a perfect world, standard commands would be standardized across the fleet, but small differences evolve between ships, often driven by personal preferences of commanding officers. For example, the order to stop engines differs from ship to ship. A commanding officer who places a premium on brevity may require the shiphandling team to use the command "All stop," but another commanding officer, preferring a certain communicating rhythm, may require the phrase "All engines stop." This reality is no different on this ship class, so shiphandlers reporting to them face the same challenge they would if reporting to propeller-driven ships:

once on board, consult the commanding officer's standing orders for guidance—and conform quickly.

The list of standard commands throughout this chapter reflects the traditions of the *Watch Officer's Guide*, the definitive work of Captain Barber, and time-tested procedures on board this ship class.[1]

Thrust Changes

Changes in thrust, whether in pierwork or open-ocean cruising, should always be made clearly and concisely. In this regard, a reminder on the terminology used for throttles is important. The written form throughout this book indicates throttle positions with the letter T. For example, if the throttle is set to position 3, it is written as "T3." This letter, however, adds little to standard commands, with their premium on brevity, so it is notably absent in the standard commands in this chapter.

In their most basic form, thrust changes, as noted in an earlier chapter, are ordered using these throttle settings. The throttles are numbered from 1 to 10 ahead and from 1 to 10 astern, with 0 indicating "stop." The throttles move smoothly across this range, so the throttles can be positioned between whole number settings. Because the waterjets use a reversing plate to redirect water discharge both forward and astern, placing the throttle at position 0 adjusts the plate to direct water in both directions at once to seek zero resultant thrust. "Ahead T1" will discharge more water astern than ahead, and, of course, "back T1" will discharge more water ahead than astern.

Between T0 and T2, the reversing plate adjusts in increments depicted in percentages, much like controllable/reversible-pitch indications on certain propeller-driven ships. When the waterjet indicator shows that it is 40 percent ahead, the reversing plate is 40 percent open in the ahead direction. Similarly, 70 percent back means that the reversing plate is 70 percent open in the astern direction. The waterjet is fully open around T2.5. Once the waterjet is fully open, subsequent increases in the throttle position will increase shaft rpm.

This sequential function of the waterjet—reversing-plate angle changes up to T2.5, followed by rpm increases above that—implies that the shiphandler can exercise precise control over thrust by either changing reversing-plate angles or rpm, but this capability has practical limitations. It will become apparent in Chapter Five, on pierwork, and Chapter Seven, on special evolutions, that this capability has diminishing value in dynamic environments with varying uncontrollable forces. Still, to ensure that the shiphandler understands the full spectrum of options available, the examples below demonstrate how the conning officer can order reversing-plate angles and shaft rpm.

It is also possible to control thrust by ordering specific ship speeds. Controlling the ship in this manner requires the helm/lee helm to use a chart that correlates throttle positions with average speeds. The reason that we refer to "average" speeds is that they are derived by driving the ship on multiple reciprocal courses. Following seas or current at one throttle position will yield faster speeds, and head-on seas or current at the same throttle position will result in slower speeds. Additionally, the ship's displacement on any given day will affect the speed produced by these throttle settings. Therefore, the speed chart gives the shiphandler an estimate of what can be expected; if a specific speed is necessary, throttle adjustments may be required.

Finally, a conning officer respecting the premium on brevity can make standard commands even more concise by accounting for the engineering plant lineup. For example, if only the starboard diesel is on line, the conning officer can simply order "Ahead three." Table 4-1 lists examples of standard commands for changing thrust with the waterjets.

Directional Changes

Since waterjets provide both thrust and directional control, shiphandlers often treat them differently depending on their employment. During docking and undocking, the waterjets are adjusted independently to produce resultant thrust vectors that move the

TABLE 4-1 Example Standard Commands for Changing Waterjet Thrust

ORDER	MEANING
Port ahead one.	Adjust port throttle to ahead T1.
Starboard back one.	Adjust starboard throttle to back T1.
Port ahead two, starboard back two.	Simultaneously adjust the port throttle to ahead T2 and the starboard throttle to back T2.
Port stop.	Adjust port throttle to T0.
All stop.	Simultaneously adjust port and starboard throttles to T0.
Port ahead for 115 rpm.	Increase the throttle ahead until the shaft rpm indicates 115 rpm.
Starboard back 30 percent.	Increase the throttle astern until the reversing plate position indicates 30 percent.
All engines ahead for ten knots.	Adjust throttles for all online engines to the position indicated on the speed chart for ten knots.
How are my engines?	Report the angle and thrust for each waterjet (for example, "Toed-out fifteen degrees, port back one, starboard ahead one").

ship in an intended direction. Once clear of the pier, the waterjets are adjusted more like rudders, both slewed in one direction to move the stern in the opposing direction. For example, if the conning officer wants to turn the ship to starboard, both waterjets are slewed to the right of centerline. This action will push the stern to port and pivot the bow to starboard. Waterjets turn the ship in a physically different way than rudders, which as noted above employ hydrofoil surfaces to produce lift to move the stern, but the similarities inherent in moving the stern to pivot the ship on a new heading are enough to refer to them like rudders.[2] As such, the standard commands for turning the ship with the waterjets not only resemble the standard commands for propeller-driven ships but even employ the term "rudder."

For the *Independence* variant, the practice of slewing the waterjets to port and starboard to turn the ship ceases when the ship achieves higher speeds. At speeds greater than about fifteen knots, the ship's control system stops moving the waterjets and engages rudders to shift the stern left and right. Since the rudders are significantly smaller than those found on cruisers and destroyers, they require a high rate of water flow across their surfaces to produce enough lift to move the stern. The operator interface is still the same, with the conning officer or helm/lee helm physically moving the combinators, and the control system determines whether to move the waterjets or rudders. Despite all this, using the term "rudder" in standard commands, whether moving the stern with waterjets or rudders, provides standardization and simplicity. When maneuvering this ship on the margins of the waterjet/rudder threshold, determining how the ship's control system is maneuvering the ship is not of much value to the shiphandler. It is important to anticipate how the ship's control system will employ the controllable forces, but the shiphandler should not spend time considering it before giving an order. When the ship is cruising at twelve, fifteen, or eighteen knots and it is time to turn the ship, turn the ship; the control system will know how to do it.

Chapter Three, on waterjet vector management, discussed the toe-in/toe-out method, wherein the order to "toe-out" or "toe-in" directs the helm/lee helm to maneuver both waterjets simultaneously to the same angle; the waterjets are considered either toed-out, toed-in, or centered. Again, the terminology of "toeing" the waterjets is drawn from the analogy of a foot pivoting on its heel, in toeing-out the waterjets is analogous to pivoting both feet outside of center, and toeing-in would resemble pivoting both inside of center. It is feasible, however, to move the waterjets individually, which would require the conning officer to specify which waterjet should be slewed to what angle.

One synonym should be noted in connection with waterjets. The final component of the drive train is referred to as the bucket, which also contains the reversing plate for changing forward/aft thrust direction. While the bucket is a specific subcomponent of the waterjet system, the terms bucket and waterjet are considered synonymous when giving shiphandling orders. Of course, no commanding officer could be faulted for mandating one or the other, by way of enforcing the second principle of standard commands discussed above. Table 4-2 provides examples of standard commands for directional changes.

TABLE 4-2 Example Standard Commands for Directional Changes

ORDER	MEANING
Toe-out fifteen degrees.	Adjust both the port and starboard waterjets fifteen degrees outside of centerline, with the port waterjet left fifteen degrees and the starboard waterjet right fifteen degrees.
Toe-in ten degrees.	Adjust both the port and starboard waterjets ten degrees inside of centerline, with the port waterjet right ten degrees and the starboard waterjet left ten degrees.
Center buckets (or waterjets).	Adjust both the port and starboard waterjets to centerline, indicating zero degrees.
Rudder amidships.	(Similar to "center buckets" but generally used in situations other than pierwork.) Adjust both waterjets to centerline, indicating zero degrees.
Right ten degrees rudder.	Adjust both the port and starboard waterjets to ten degrees right of centerline. If only one maneuverable waterjet is online, adjust only the online waterjet.
Left twenty degrees rudder.	Adjust both the port and starboard waterjets to twenty degrees left of centerline. If only one maneuverable waterjet is online, adjust only the online waterjet.
Shift your rudder.	(Given only when the waterjets are adjusted at the same angle off centerline.) Slew the waterjets to the same angle on the other side of centerline. (For example, if the waterjets are left fifteen degrees, shift both waterjets to right fifteen degrees.)
Increase your rudder to right thirty degrees.	(Given to increase the rate of turn or stern swing.) Increase the waterjet angle to the ordered position (in this example thirty degrees right of centerline).
Ease your rudder to left five degrees.	(Given to decrease the rate of turn or stern swing.) Decrease the waterjet angle to the ordered position (in this example five degrees left).
Set autopilot for 180.	Engage autopilot if not already so and set it for course 180 degrees true.

(continued)

TABLE **4-2** **Example Standard Commands for Directional Changes**
(continued)

ORDER	MEANING
Come right, steer course 270.	(Given when the course change is less than ten degrees.) Alter course to starboard and steady on course 270 degrees true. If autopilot is in use, set autopilot for 270 degrees true.
Left fifteen degrees rudder, steady on course 090.	(Given when the course change is greater than ten degrees true.) Set the waterjet angle to fifteen degrees left of centerline and steady on course 090 degrees true. If autopilot is in use, take the rudder in hand, but once the turn is complete, set autopilot for 090 degrees true.
Steady on course 215.	(Given when the ship does not have an ordered course and the ship is pointing roughly along the intended heading.) Begin steering 215. If autopilot is engaged, set autopilot for 215.
Steady as she goes.	(Given when the ship does not have an ordered course.) Begin steering the heading when the order is issued. If autopilot is engaged, set autopilot for this course.
How are my waterjets?	(Synonymous with "How are my engines?") Report the angle *and* thrust for each waterjet. (For example, "Toed-out thirty degrees, port ahead one, starboard back one.")

Bow Thruster

As with standard commands for the waterjets, orders to control the bow thruster must be clear and concise. The bow-thruster combinator is numbered in a way similar to waterjet control; a throttle set to position 5 is said to be at "T5." Also as with waterjets, this letter adds little to standard commands, designed for brevity, so it is deliberately eliminated in the standard commands of table 4-3.

Changes to the bow thruster's power output are ordered using these throttle setting numbers, from 1 to 10 ahead, 1 to 10 astern, and 0 indicating "stop." Since the thruster can be maneuvered 360 degrees in little time, the thruster is not generally used with reverse thrust; pointing the thruster toward 090 degrees relative and then slewing it to 270 degrees relative will achieve the same

TABLE 4-3 Example Standard Commands for the Bow Thruster

ORDER	MEANING
Train thruster 090.	Adjust the bow thruster direction to 090 degrees relative.
Thruster ahead four.	Adjust the bow thruster throttle to T4.
Thruster stop.	Adjust the bow thruster throttle to T0.
How is my thruster?	Report the direction and output of the thruster. (For example, "Thruster trained 270, ahead six.")

result as reversing the thrust direction, and the shiphandling team is not then required to keep track of whether the thruster is ahead or back. Keeping the thruster ahead brings a certain simplicity to the shiphandling evolution—every member of the team knows that wherever the thruster is pointed, that is the direction in which it will pull the bow. In addition, the thruster is more effective ahead than back, so keeping the thruster ahead ensures maximum control on the bow. Table 4-3 provides examples of standard commands for the bow thruster.

Tug Control

In general, tugs are controlled by the harbor pilot, but this highly maneuverable ship class has caused the commanding officers to rethink the need for pilots in home port, particularly during the less challenging task of undocking. The specifics for using tugs, including number and positioning, will be discussed in detail in Chapter Five, but it is worth noting here that there are many ways to be creative with tugs when maneuvering alongside the pier. The following commands focus on the most common functions in tug work.

Although the most advanced shiphandlers can accomplish extraordinary maneuvers with imaginative tug combinations, doing so requires close coordination with the tug captains to reduce the possibility of communication errors. As is the case with standard commands for shipboard controllable forces, simple and repeatable maneuvers are generally preferable to complex combinations. The commands below and the subsequent discussion in Chapter Five center on tug movements toward and away from the ship, focusing the tug's efforts on its greatest contribution to pierwork—immediate and powerful lateral force.

This simplicity is particularly important when using the tugs as a "safety net," merely tied to the ship in the event of an emergency. The tug may be needed in case of an engineering casualty that causes the ship to lose shiphandling control or if uncontrollable forces are too great to overcome with the ship's controllable forces alone. In such emergencies, whether or not the tugs are under the control of the pilot, the quickest way to put them to work is through clear and concise direction. Tug captains and harbor pilots will react quickly to straightforward language in any situation, but particularly during emergencies.

As discussed in Chapter Two, on controllable and uncontrollable forces, tugs can have either one or two shafts. The most capable tugs, again, are Z-drive craft with two shafts that can direct their thrust 360 degrees. The thrust of these tugs is so powerful compared to the ship's lightweight design that it produces a thrust-to-weight ratio that can quickly become hazardous if not carefully controlled. In effect, the ship's safety net can unexpectedly become its worst hazard; the tug's power must be employed judiciously.

Tug power is commonly controlled with the commands *Easy* for one-third power, *Half* for two-thirds power, and *Full* for 100 percent output. One way to dampen the tug's power is to call for even finer control at its least powerful output setting. A call for "Easy one" will cut the tug's power in half by directing it to use only one shaft and at one-third power. If easy one is still providing too much power, the tug can be asked for "dead slow" to request the minimum power possible. Of course, this measure may not be necessary for single-screw tugs, which are less powerful by design.

Additionally, there is no internationally accepted terminology for tug standard commands, so when out of home port, whether working through a harbor pilot or directly with the tug captains, it is worthwhile to review the locally accepted terminology before

TABLE **4-4**   Example Standard Commands for Tugs

ORDER	MEANING
Make up single head line on the bow.	Pass a single line from the tug to the forecastle, and make the line fast for the tug to pull.
Cast off.	Remove the tug's line from the ship's bitts and return to the tug.
Away dead slow (or, pull dead slow).	Pull on the ship at a ninety-degree angle with minimum power.
Toward dead slow (or, push dead slow).	Push on the ship at a ninety-degree angle with minimum power.
Away easy one (or, pull easy one).	Pull on the ship at a ninety-degree angle with one-third power using one shaft.
Toward easy two (or, push easy two).	Push on the ship at a ninety-degree angle with one-third power using two shafts.

pierwork begins. For example, some tug captains prefer to be directed toward or away, with respect to the ship. Others like to be directed to pull or push, and still others prefer to receive a direction of port or starboard. Some even like to hear a combination, such as "Away to port" or "Push to starboard." In any case, these nuances should be understood long before an emergency develops, at which point the premium on clarity is the highest. Table 4-4 provides examples of standard commands for tugs.

CONCLUSION

In the end, the standard commands presented in this chapter are not much different from the time-tested commands used throughout the fleet on propeller-driven ships. These commands simply employ a language that relates more clearly to the waterjets of these ships. The use of tugs or thrusters to control the bow is certainly not unique to this ship class, but the combination of propulsion and rudders into a single waterjet has upended conventional shiphandling terminology.

But if the innovative design of this ship class has overturned some long-standing methods of shiphandling, the benefits of standard commands endure. From the situational awareness among the shiphandling team to the critical safety net provided by more advanced shiphandlers, vocalizing engine and waterjet configurations before enacting them can save the ship from allision or collision; also, formalizing the exchange of orders ensures that the process is there when you need it most. Positive control of the ship begins with clear communication, and clear communications begin with standard language.

CHAPTER FIVE

Pierwork

Shiphandling is a skill that requires regular practice to achieve and maintain proficiency. In a cruel twist, however, frequent shiphandling can increase the potential for an allision, as complacency grows each time the ship docks and undocks without incident. Familiarity breeds this complacency, and complacency is the shiphandler's worst enemy during pierwork, particularly under the most benign conditions. When uncontrollable forces are less cooperative, the shiphandler is more alert; when the wind is calm and the current is slack, however, shiphandlers can easily let down their guard. Naval officers study and practice shiphandling in order to achieve a level of proficiency that makes docking and undocking relatively simple, but maneuvering the ship alongside a pier should never be considered business as usual.

One of the best ways to ameliorate the risks associated with pierwork is to develop and practice *simple, repeatable methods* that are effective in any environmental condition. Whether the wind is calm or gusting fifteen knots, the methods employed to maneuver the ship safely alongside the pier should be relatively uniform and always straightforward. This chapter aims to present these methods in a clear, logical way, beginning with a discussion of pierwork fundamentals that apply in all environmental conditions. It will then walk through the basic evolutions of docking and undocking the ship, including focused explorations of linehandling and the use of tugs and pilots. These sections will lead to more advanced concepts of moving the pivot point on demand, dynamic bow-thruster management, and pierwork in casualty configurations.

PIERWORK FUNDAMENTALS

The most important aspect of docking and undocking this ship class is controlling the ship's lateral motion. This ship's ability to move laterally is one of its greatest assets, but if this movement is not carefully controlled, it can become the ship's greatest hazard, particularly when docking, since it will be moving directly toward the pier. This capability is useful only if the shiphandler knows how to move the ship in the right direction and, more importantly, how to stop the ship when it is moving in the wrong direction. Fortunately, the waterjets on this ship class provide the means to do just that.

Below are several fundamentals of pierwork worth remembering when docking and undocking the ship.

Watch the "Three Gauges for Pierwork"

Walking the ship laterally is fairly simple to understand when looking at a top-down illustration, because then it is easy to envision the ship moving sideways. It is more difficult, however, to stand on the bridge of the ship and try to discern whether the ship is actually moving laterally. High-tech sensors and displays provide a great deal of information to the conning officer, but like most data, increasing amounts of information have diminishing value. Without a method to filter the data or reduce it to only relevant indicators, the conning officer can become distracted by information that is relatively meaningless. Put another way, any time the conning officer spends looking at the *least* significant information is time not spent focusing on what is *most* significant.

So, when trying to discern if the ship is walking laterally, what information is the most important? Without the benefit of a top-down perspective, how does the conning officer know that the ship is moving sideways? The conning officers knows whether or not the ship is in fact walking laterally by simply watching three gauges—the ship's heading, longitudinal speed, and lateral closure relative to the pier.

Heading

As long as the conning officer knows the pier heading, the ship's heading will indicate whether the ship remains parallel to it. If the ship is pivoting to port, the heading will decrease; and if the ship is pivoting to starboard, the heading will increase. Any number of motions will cause the ship to deviate from parallel. For example, a pivot to port may be caused by the bow outpacing the stern as both move away from the pier, or the bow may be moving to port as the stern is swinging to starboard. The heading alone does not show how the ship is pivoting, but it will be the first indication that the bow and stern are not moving together at the same lateral speed. The reality of shiphandling in a dynamic environment of uncontrollable forces makes it unlikely that the conning officer will ever maintain a perfect heading throughout a lateral walk, so it is reasonable to expect a ship that is laterally walking to vary plus or minus five degrees from the pier heading.

Longitudinal Speed

Longitudinal speed will indicate whether the ship is sliding ahead or astern. Any considerable longitudinal movement means that the ship is not walking laterally but instead is crabbing diagonally.[1] As with the ship's heading, uncontrollable forces make it unlikely that a ship will be able to maintain zero knots ahead or astern throughout the maneuver; it can be expected that the ship will move forward or backward at one-tenth of a knot. It is important to remember, however, that even low speeds result in considerable distances over time, so the shiphandler should be mindful of the *average* longitudinal speed throughout the evolution. For example, if the longitudinal motion bounces between 0.0 knots and 0.1 knots throughout the undocking, the ship is steadily moving ahead in the slip. If the speed varies evenly between minus 0.1 knots (i.e., astern) and 0.1 knots, the ship is remaining practically in place.

Of course, the best way to verify the ship's longitudinal position is to find a marker on the pier—a set of bits, a bollard, a shore-power bunker—that lies roughly in line with the bridge wing. If the ship is moving forward or aft, the ship's speed will indicate the severity of the problem, but the marker will serve as ground truth for the ship's position. Given time, novice shiphandlers will learn to trust their eye to judge longitudinal speed more than electronic instrumentation.

Lateral Closure

Best observed from the bridge wing, lateral closure relative to the pier refers to whether the ship is opening or closing. Distance can be estimated with a single look, but closure can be determined only by checking repeatedly. Simply put, "ten feet and opening" is a starkly different shiphandling situation than "ten feet and closing," and the only way to know confidently whether the ship is moving toward or away from the pier is to check visually the distance between them, look again, and then look again. Since two-thirds of the ship is aft of the bridge, the conning officer should spend more time looking aft than forward, but it is important to check along the entire length of the ship. Any combination of the bow or stern opening, closing, or holding steady is possible.

Individually, each of these gauges tells the conning officer about one aspect of the ship's movement, but together they, and they *by themselves*, will indicate if the ship is walking laterally. Consider a ship with the pier on the starboard side: if the ship remains on the pier heading, the longitudinal speed is zero knots, and the ship is opening (moving away from) the pier, then it is laterally walking to port. If the ship remains on the pier heading, the longitudinal speed is zero knots, and the ship is closing the pier, then it is walking laterally to starboard. Again—and this point cannot be stressed enough—no other gauges are required to know whether or not the ship is walking laterally, so the conning officer should practice rotating between them continuously.

Table 5-1 lists several ship movements indicated by the three gauges for pierwork, given that the pier is on the starboard side. The goal is to use the information from these three gauges to envision the ship's movement relative to the pier.

TABLE 5-1   Example Ship Movements Indicated by the Three Gauges for Pierwork, Given a Pier on the Starboard Side

HEADING	SPEED	LATERAL CLOSURE	SHIP MOVEMENT
Steady	0 knots	Bow and stern closing	Starboard lateral walk
Steady	0 knots	Bow and stern opening	Port lateral walk
Decreasing	0 knots	Bow opening/stern closing	Port twist
Decreasing	0 knots	Bow steady/stern closing	Pivoting on the bow to port
Decreasing	0 knots	Bow opening/stern steady	Pivoting on the stern to port
Increasing	0 knots	Bow closing/stern opening	Starboard twist
Increasing	0 knots	Bow steady/stern opening	Pivoting on the bow to starboard
Increasing	0 knots	Bow closing/stern steady	Pivoting on the stern to starboard
Steady	1 knot ahead	Bow and stern closing	Starboard-forward diagonal crab
Steady	1 knot astern	Bow and stern opening	Port-aft diagonal crab

Align Controllable Forces against Uncontrollable Forces

Recalling the discussion in Chapter Two on controllable and uncontrollable forces, it is clear that both categories of forces have roles in the ship's movement. The uncontrollable forces push the ship wherever they will, and the controllable forces enable the shiphandler to manage their effects. To be sure, the uncontrollable forces do not always push the ship in the wrong direction, but it is very rare that they push it *entirely* in the right direction.

As a starting point, the shiphandler should align the controllable forces opposite to the uncontrollable forces whenever the ship is moving toward a pier hazard. For example, if the ship is docking with its starboard side to the pier and with an onsetting wind (i.e., from port to starboard), the shiphandler should begin the maneuver with the waterjets and thruster (or tug) set for a port walk. The uncontrollable force will blow the ship down onto the pier, and the shiphandler will use the controllable forces as a brake as the ship is blown into place. Were the conning officer to point the controllable forces in the same direction as the uncontrollable forces, the effect would be doubled, accelerating the ship toward the pier hazard. The uncontrollable forces are often insufficient to move the ship, but as a rule of thumb in such a situation, beginning the docking evolution with the controllable forces pointed into the wind will ensure that the docking begins slowly and safely.

Trust the Toe-In/Toe-Out Method

Chapter Three, on waterjet vector management, introduced the shiphandler to the toe-in/toe-out method.[2] Although that discussion showed that there are many ways to combine waterjets to achieve similar effects, the simplicity of the toe-in/toe-out method itself is hard to overlook. More importantly, though, this method provides maximum control over the stern at all times. Crenshaw's first rule of shiphandling is to always keep the stern clear of danger, because if the propellers and rudders are damaged, the ship is paralyzed.[3] This rule is just as important today; the waterjets must be protected with equal vigor. The shiphandler should seek the simplest and most effective method to maintain control of the stern.

The toe-in/toe-out method provides this control; by simply altering the angle of the waterjets, the conning officer can either accelerate or decelerate the stern's lateral movement. Additionally, this effect can be realized without changing the engine's speed or longitudinal direction. When toeing-out to walk the ship laterally, increasing the toe-out angle will accelerate the stern faster into the walk. If the stern is moving too quickly toward danger, centering the waterjets will "tap the brakes," and toeing-in will "slam the brakes." These angle changes take just seconds to apply, so the shiphandler can quickly and precisely maintain control over the stern.

Because this method is so effective in controlling the stern, its usefulness is not restricted to moving the stern during docking or undocking. Handling the ship while adjacent to piers, including maneuvers where the ship is moving directly ahead or astern, always requires command of the stern's movement. For example, when the ship has cleared the pier during an undocking evolution, the conning officer will aim to keep the ship in the center of the slip while moving toward the channel. It is certainly possible to maneuver with both waterjets ahead, or astern if the ship is backing into the channel, but if the toe-in/toe-out method had been so effective in controlling the stern while laterally walking away from the pier, why surrender that capability while still adjacent to the pier hazard? Maximum control over the stern is critical when uncontrollable forces are trying to push the ship back down on the pier.

The three gauges for pierwork—heading, longitudinal speed, and lateral closure—can also assist the shiphandler in transiting along the center of the slip. The ship's heading should still remain within a few degrees of pier heading to ensure that it lies nearly parallel to that hazard and therefore longitudinally aligned relative to the center of the slip. Lateral closure to the pier should be kept to zero, to ensure that the ship is neither moving toward the pier nor crossing into the other side of the slip. Finally, if the bow is pointed toward the channel, the longitudinal speed can simply be allowed to build up ahead. This is easily achieved by using the ahead engine to overtake the astern engine. For example, if the port engine is back T1 and the starboard engine is ahead T1, increasing the starboard ahead engine to T2 will produce a slight resultant force vector in the ahead direction. Increasing the starboard ahead engine to T3 will build even more headway, all while maintaining lateral stern control provided by the traction of the port astern engine.

It may seem counterintuitive to leave an engine backing while driving ahead, but this is the essence of waterjet vector management described in Chapter Three. The direction of the individual waterjets matters far less than the resultant force vector, particularly when the reverse direction of the waterjet is providing an important benefit. In this case, the astern engine provides the shiphandler with the precise lateral stern control inherent in the toe-in/toe-out method as the ship moves steadily out of the slip.

Minimize Thrust Direction Changes

Whenever a ship, whether in this ship class or a propeller-driven ship, reverses engine directions, it experiences an awkward moment caused by several factors. First, reversing the engine from ahead to astern or from astern to ahead requires the propulsion to move through a point of zero thrust, at which the ship is momentarily at the mercy of the uncontrollable forces. Second, once the engine is producing the desired thrust, the thrust in the new direction takes some time to create an "equal and opposite reaction," during which, again, the ship is subject to uncontrollable forces. Finally, human error in maneuvering the throttles, combined with the normal latency in engineering control systems, yields mixed results even when trying diligently to time engine reversal exactly. Taken together, these factors mean that it is very difficult to reverse the lateral force vector precisely by changing engine direction.

Consider how this effect would look during pierwork, for example, if the stern were moving laterally to starboard toward the pier, with the waterjets toed-out fifteen degrees, the port engine ahead T1, and the starboard engine back T1. Let's say that the stern is moving too quickly toward the pier due to an onsetting wind, so the conning officer, trying to brake the stern, decides to reverse the engines, ordering the port engine back T1 and the starboard engine ahead T1. This configuration would nominally cause the stern to begin moving laterally to port, but the factors listed above will complicate the transition. First, the movement through the zero-thrust point on each waterjet will create a moment in which the ship is being carried along by the onsetting wind with no controllable force to combat the movement. Even when the waterjet reaches its desired thrust, this moment persists while the action of the waterjet is generating an equal and opposite reaction on the stern. Finally, the starboard ahead waterjet may achieve thrust just a fraction of a minute earlier than the port astern waterjet, or vice versa, which could produce an undesired longitudinal force vector. Of course, all this is assuming that the helm/lee helm positions the throttles in the exact reverse of the original configuration.

Now, consider how this maneuver would appear if the conning officer simply reversed the waterjet angle from toed-out fifteen degrees to toed-in fifteen degrees. The waterjets would never go through a zero-thrust point, and, since the thrust direction would remain constant, there would be no opportunity for one waterjet to gain traction before the other did. There will still be a slight delay as the action produces a lateral reaction, but the controllable forces are *immediately* countering the effect of the onsetting wind working against the ship.

The point of highlighting these factors is that reversing the engine thrust during pierwork creates unnecessary uncertainty that could be avoided by simply reversing the toe angle instead. This advantage has more applications than simply changing the direction of a lateral walk. By understanding that the stern can move either left or right with the engines remaining in the same longitudinal direction, the shiphandler should be able to see that there is very little advantage in reversing the engines during pierwork. Furthermore, this should cause an experienced water-jet shiphandler to question any allegiance to a toed-out walk or a toed-in twist, since the ship can execute a toed-in walk or a toed-out twist just as easily and without the uncertainty of reversing thrust direction. Once having selected which engine is ahead or astern, keeping the same thrust direction throughout the day's pierwork is the safest approach.

To appreciate the holistic nature of this method, consider an entire docking evolution that includes both twisting and walking, from entering the slip to mooring starboard side toward the pier. For example, as illustrated in figure 5-1, a ship moving down the channel with the piers on the port side may need to twist to port and walk to starboard. If the conning officer chooses a toed-in configuration to twist into the slip (position A), with the starboard engine ahead and the port engine back, the starboard engine

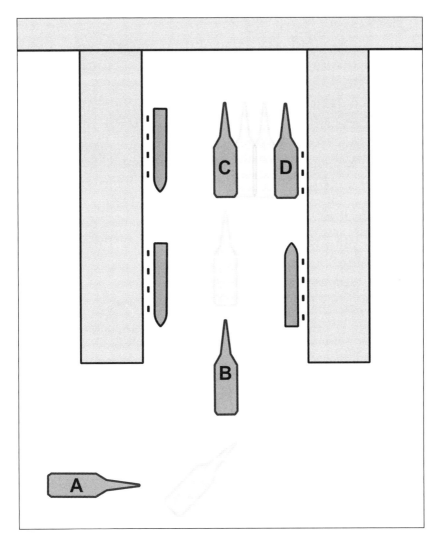

FIGURE 5-1   Full Docking Maneuver, from Channel to Berth, Using Only the
             Toed-In Configuration

should remain ahead and the port engine back until moored. Once
the ship is in the center of the slip (position B), the starboard ahead
engine should be increased to overcome the port astern engine and
produce a resultant force vector ahead. When the ship is aligned
adjacent to the mooring position (position C), the headway should

be checked with increased power to the port astern engine, and when ready to walk laterally toward the pier, the conning officer should use a toed-in configuration to move the ship to its berth (position D).

The advanced shiphandler is able to envision the entire ship-handling evolution from beginning to end and determine in advance thrust direction during the most critical maneuvers. In the above example, if the lateral walk toward the pier is considered to be the riskiest maneuver because of certain uncontrollable forces and the conning officer prefers a toed-out configuration for that stage, the entire docking evolution could be reverse engineered to ensure that engines are already in the preferred direction when commencing the toed-out walk. If the advanced shiphandler desired a toed-out walk for the maneuver depicted in figure 5-1, the ship would employ a toed-out configuration to twist the ship into the slip (position A to B), with the starboard engine ahead and the port engine back. Once in position adjacent to the pier (position C), the waterjets would be aligned for a toed-out lateral walk to starboard (position D).

Control Momentum with the "By-Half" Technique

When maneuvering during pierwork, momentum can be the ship-handler's worst enemy, specifically when it is in the wrong direction. For example, if the stern is quickly swinging toward the pier when docking the ship, the most pressing task is to stop that lateral movement immediately. The novice shiphandler will likely surmise that power has been applied in the wrong direction and then seek "the right amount of power in the right direction" that should have been applied to begin with. This problem-solving approach will cause the stern slowly to reduce speed toward the pier hazard before coming to a halt and reversing direction, but only if there is enough room to reverse direction before an allision occurs. Instead, when the ship is moving in the wrong direction, the conning officer should create the maximum resultant force vector that is safe during pierwork to stop this momentum. Once the ship is no longer moving in the wrong direction, the conning officer can seek the right amount of power to move the ship in the right direction.

A simple way to execute this maneuver is the *by-half technique*.[4] The shiphandler uses power that is clearly greater than would be required to move the ship were it standing still. Once momentum in the wrong direction is halted the conning officer reduces this power by half, and when momentum in the right direction begins developing reduces it by half again. The conning officer continues this sequence until just enough power is being applied to move the ship in the right direction. This technique can be applied with any of the propulsive controllable forces—engine power, waterjet angle, thruster power, or tug power—to save the ship from pending danger.

Engine Power

If the ship is using the toe-in/toe-out method with the starboard engine ahead T1 and the port engine back T1 and unexpectedly builds sternway toward a quay wall, the conning officer could quickly apply ahead propulsion by increasing the ahead waterjet to T4, then reducing it to T2, and then reducing it again to T1.

Waterjet Angle

Under the above engine configuration for undocking, if the stern began suddenly moving starboard toward the pier, the conning officer could quickly apply a port lateral force by toeing-out thirty degrees, then reducing the toed-out angle to fifteen degrees, then again to 7.5 degrees.

Thruster Power

Again under the same configuration for undocking, if the bow began moving starboard toward the pier, the conning officer could quickly apply a port lateral force on the bow with the thruster trained to 270 degrees relative by increasing the power to T8, then reducing it to T4, then again to T2.

Tugs

Without a thruster available in this scenario, the same technique could be used to pull the bow away quickly. The conning officer could order the tug to pull away to port "Easy two," then reduce it to "Easy one," and then "Dead slow."[5]

Avoid Overpower

When maneuvering in close quarters to the pier, excessive power can create momentum very quickly, and the shiphandler can rapidly lose control of the ship. Using too little power will cause the ship to react sluggishly, but too much can produce a situation that is far more dangerous. Prior to undocking, the shiphandler should evaluate the uncontrollable forces and determine how much power is required to push the ship off the pier, but the difference in consequences between "underpower" and "overpower" should cause the shiphandler to err on the side of less power.

Throttle settings between T0 and T2 will adjust the bucket reversing plate, slightly opening and closing it to change how much water is discharged in the ahead and astern direction. Once the throttle reaches about T2.5, the bucket is fully open, meaning that 100 percent of the water is discharging astern, and any subsequent increases in throttle settings will adjust shaft rpm while the reversing plate remains open 100 percent. For shiphandlers experienced on controllable/reversible-pitch ships, this relationship should be familiar, since propeller pitch and shaft rpm operate in the same manner. This ship class differs from those ships in that its weight is so much less. Its thrust-to-weight ratio is considerably greater, so less power is needed to move it around. Consequently, the shiphandler can push the ship away from the pier with the throttles set to as little as T1 on both engines.

To prevent overpower from occurring, an incremental approach to engine power is warranted. For example, when undocking the ship with an onsetting wind using the toe-in/toe-out method, the conning officer should start with both engines at T1. If the stern does not budge even with both waterjets toed-out thirty degrees, the conning officer will know that more power is required to overcome the onsetting wind. After increasing power to T2 on both engines, the shiphandler should watch how the stern reacts. If the stern still remains in place, the conning officer can increase power on both engines again but should be aware that the difference between T1 and T2 is not the same as between T2 and T3, because the latter includes the transition from altering the reversing plate only to altering shaft rpm. This power transition is not linear. Of course, any time that momentum toward the pier hazard develops, quick action is called for, and the shiphandler should employ the by-half technique described above to regain control of the ship.

Beware Baseline Creep

The toe-in/toe-out method centers on combining two opposing waterjets in such a way that the ahead and astern force vectors cancel out to allow the ship to walk laterally. Conceptually, this cancelation is achievable with both engines at T1, T2, T3, or any equal but opposing combination. The above discussion on avoiding overpower should lead the shiphandler to appreciate that the safest combination is T1 on both engines, but in the example described above sometimes T2 or T3 is required to overcome strong winds. The minimum throttle setting that proves capable of moving the ship effectively should be considered that day's *baseline* for the toe-in/toe-out method.

Once this baseline is determined, the shiphandler should be careful not to change it inadvertently without making a deliberate decision that more power is needed. *Baseline creep* can occur innocently enough as the conning officer tries to control longitudinal motion. For example, let's say that the baseline for a day with light winds is T1 on both engines and that while walking the ship laterally to port during undocking, the ship begins building sternway. The conning officer increases power on the ahead engine to T2 to check that movement, but if the ship begins building headway and the conning officer increases power on the astern engine to T2 to check that motion, the ship is now at a new baseline of T2. One can see that repeating this sequence can result in baseline creep, and more power can be employed than is necessary simply because the shiphandler did not control the baseline. Unless it has been determined that more power is required to move the ship effectively, the conning officer should aim to return to the established baseline to prevent an overpower situation.

UNDOCKING

Between undocking and docking the ship, undocking is by far the easier shiphandling evolution. This is not to say that undocking the ship is a risk-free maneuver, but since the ship is accelerating away from the hazard, the shiphandler does not need to exercise as much control over the lateral speed of the ship. As long as the ship is moving toward the center of the slip in a controlled manner, this evolution is fairly straightforward.

One significant difference between the *Independence* and *Freedom* variants should be noted when considering how to control contact with pier fenders during undocking.[6] The physical design of the hull requires the conning officer to handle each variant differently. The *Freedom* variant has a traditional hull design that roughly resembles that of a frigate or corvette. Even though its draft is far less, resulting in a flatter underwater hull along the keel, the sides of the hull are not extraordinary different; so, a shiphandler with experience on other small types—frigates, destroyers, cruisers—will be able to manage hull contact with the fenders in the familiar way. Specifically, it is possible to roll on the fender, where the conning officer kicks out the stern well ahead of the bow, as long as there are sufficient fenders along the forward end of the ship to prevent hull contact with the pilings. In contrast to other types of ships with hull-mounted sonars that must be protected, the *Freedom* variant can easily execute this maneuver.

The *Independence* variant, however, does not have a traditional hull. The trimaran design often requires the shiphandler to think of the ship as having three hulls, and this is particularly relevant during pierwork. This variant cannot roll on the fender, because the amah—the outer hull in contact with the fenders—ends just under the bridge, about one-third of the ship's total length from the bow. The leading edge of the amah is shaped to a knife edge to cut through waves at sea, but in port the hull's incongruous design prevents the ship from rolling smoothly along the entire length of the bow section as the *Freedom* variant can. This leading edge acts like a very sharp shoulder, and any attempt to pivot on it will yield results that are difficult to predict, which is particularly hazardous when the pier is lined with small fenders. For example, the amah's leading edge can insert itself between two fenders and directly contact the pier. Alternately, if the ship touches a small fender with the leading edge rather than the flat side of the amah, the fender may pivot independently and expose the hull to the pier pilings. The ship does not need to remain precisely flat when making contact with the fender, but swinging the stern out more than about ten degrees increases the probability of undesired hull interaction. The best way to prevent this type of allision is to undock the *Independence* variant by laterally walking away on the pier heading.

Long before getting under way, the conning officer should carefully examine the uncontrollable forces and consider how they will affect the ship once the mooring lines are removed. This evaluation should begin the previous day when the predicted weather conditions are briefed, but the shiphandler should make the final determination just moments before assuming the watch on the bridge. Once the sea and anchor detail is set, the conning officer is well advised to invest five minutes on the forecastle and flight deck to sense the elements, such as by watching the flags on adjacent ships to ascertain wind direction and examining the channel buoys to verify the direction of the current.

Noting the wind direction at both the bow and stern is particularly important, because it is not uncommon to have wind direction vary between the bow and stern. Weather forecasts do not predict this effect, and relying solely on the anemometer will indicate only the relative wind at one point on the ship. Surrounding structures in the basin, such as adjacent ships and pierside buildings, can affect the uncontrollable forces. For example, when the ship is moored across the pier from a significantly larger vessel, the superstructure of that ship can cause a swirling eddy of wind, one so pronounced that the relative wind on the bow may be on the opposite side from that on the stern. The wind could in fact be pushing the bow off the pier at the same time that it is holding the stern onto the pier. Modern weather predictions are more accurate than ever before, but nothing replaces real-time observations on the points of the ship where the uncontrollable forces really matter.

With a firm grasp for the observed uncontrollable forces, the conning officer can refine the plan to use the controllable forces to push the ship off the pier. A strong offsetting wind should make for a fairly simple evolution; simply taking in the mooring lines might be sufficient to free the ship from the pier, as the wind pushes the ship away. An extraordinarily strong onsetting wind, however, could require more power than the ship's engines can safely provide, so the conning officer may need to employ tug power to pull the ship free from the pier. With the relative wind on opposite sides at the bow and stern, the wind may be sufficient to free one end of the ship, but controllable forces would be required for the other end. As discussed in Chapter One, this is where shiphandling becomes more art than science, and for the more experienced practitioners, where it can be the most fun.

In refining the shiphandling plan, the conning officer would benefit from considering the *six shiphandling actions* discussed in Chapter Three. For example, table 5-2, on the six shiphandling

TABLE 5-2 Six Shiphandling Actions for Walking the Ship Laterally to Port, with the Starboard Engine Ahead and Port Engine Back

SIX SHIPHANDLING QUESTIONS	SIX SHIPHANDLING ACTIONS
1. What makes the stern go to port?	Toe-out waterjets
2. What makes the stern go to starboard?	Toe-in waterjets
3. What makes the bow go to port?	Thruster to 270 degrees relative, or tug pulling away to port
4. What makes the bow go to starboard?	Thruster to 090 degrees relative or tug pushing toward to starboard
5. What makes the ship move ahead?	Increase the starboard ahead engine
6. What makes the ship move astern?	Increase the port backing engine

actions for laterally walking the ship to port with the starboard engine ahead and the port engine back, would be useful to have on hand for quick reference.

Additionally, the conning officer should consider the three gauges discussed above—heading, longitudinal speed, and lateral closure—and how different indications will signal the ship's actual movements. Remember that the conning officer does not benefit from a top-down view of the ship, so it is important to use these three gauges to determine how the ship is moving. Table 5-2 can serve as a template of a guide for laterally walking the ship to port and will help the conning officer determine the ship's movement while undocking.

To put these concepts into action, consider how an undocking evolution would proceed with negligible uncontrollable forces affecting the ship. Take a ship moored starboard side toward the pier with a heading of 270, bow out toward the channel. According to the six shiphandling actions in table 5-2, the initial configuration to walk the ship laterally to port would be to toe-out thirty degrees, with the starboard engine ahead T1, port engine back T1, and the thruster trained to 270 degrees relative at T4 (or a tug pulling away to port).

In an ideal environment, these actions would happen simultaneously, so that all controllable forces applied evenly. It is practically impossible to make that happen, so some prioritization is in order: waterjet angle first, engine power second, and thruster (or tug) third. The waterjets are moved first, because changing the waterjet angle without power will not affect the stern considerably. Remembering Crenshaw's first rule of shiphandling—always keep the stern clear of danger; in this case, engine power is applied to move the stern laterally away from the pier when all mooring lines are clear.[7] As soon as the stern is seen to be opening from the pier, power is applied on the bow with either the thruster or tug to achieve lateral movement to port.

At this point, the conning officer begins watching the three gauges carefully. If the heading reads 275 and is increasing with the ship opening the pier, the conning officer knows that the stern is outpacing the bow; the waterjet toe angle must be reduced. Conversely, if the heading reads 265 and decreasing with the ship opening the pier, the bow is outpacing the stern, and power on the bow should be reduced. If the ship begins sliding ahead, the conning officer knows that the starboard engine is overpowering the port engine, so the port engine should be increased to back T2. If instead the ship builds sternway, the port engine is overpowering the starboard engine, so the starboard engine must be increased to ahead T2. Finally, while the lateral closure will seem self-evident to the experienced shiphandler, the novice should remember to check this gauge continually to ensure that the entire length of the ship is moving away from the pier.

Once the ship is in the center of the slip, the conning officer should stop the lateral motion and begin moving ahead into the channel; the six shiphandling actions are reliable tools to control the ship throughout this maneuver as well. To stop the stern from moving to port, the conning officer should briefly toe-in the waterjets to stop the stern's momentum. Once the lateral movement has been stopped, the waterjets can be centered again. Controlling the lateral movement of the bow too is made easier by referencing the six shiphandling actions. Slowing the bow's lateral movement is accomplished by stopping the bow thruster pointed to 270 degrees relative, or stopping the tug pulling away to port. If the bow is moving very quickly to port, this movement can be checked by training the thruster to 090 degrees relative or directing the tug to push to starboard. Once the bow's lateral movement is halted, the ship can drive ahead on the waterjets alone. Still, just as it is recommended to keep the backing engine on for maximum control of the stern, the shiphandler is wise to continue using the bow thruster if available until the ship is clear of the slip.

Dynamic bow-thruster management is discussed in more detail later in this chapter, but for now, one basic rule of thumb is worth mentioning for the thruster: point the thruster wherever you want the bow to move. This rule may seem self-explanatory for the advanced shiphandler, but its simplicity is what makes the thruster so useful. In the example above, once the ship is centered in the slip, the conning officer wanting the bow to move forward trains the thruster to 000 degrees relative. If the bow falls off slowly to starboard, the conning officer simply has to train the thruster to 315 degrees relative to pull the bow back to the center of the slip. As will be discussed later in this chapter, the force created by training the thruster ahead adds to any headway created by the waterjets. Minimizing power on the thruster should provide sufficient control over the bow without considerably adding to the ship's overall speed.

Once the ship is centered in the slip, the six shiphandling actions provide the answer for building headway. Simply increasing the starboard ahead engine to overcome the port backing engine will allow the ship to slide forward carefully. For example, setting the starboard ahead engine to T2 while maintaining the port backing engine at T1 should build about one to two knots of headway. Increasing the starboard ahead engine to T3 will boost the ship's speed to between two and three knots of headway. As the ship moves ahead toward the channel, the conning officer can still manage the lateral movement of the stern with the toe-in/toe-out method; the six shiphandling actions still prove useful here. The shiphandler knows that toeing-out will move the stern to port and that toeing-in will move the stern to starboard.

As the ship enters the channel, keeping the bow thruster down (or tug tied on) provides a great advantage in quickly turning into the channel. Let's say that the first leg of the outbound channel is on course 000 degrees true and that a starboard turn will be required. As soon as the bow is clear of the buoy line, the conning officer can begin working the bow into the channel by directing the bow thruster to 090 degrees relative or by ordering the tug to push toward to starboard. The pivot point will have moved forward because of the ship's headway and then it will move aft owing to the lateral force applied on the bow, returning approximately to its natural position, so pushing the bow to starboard will cause the stern to pivot to port at about the same rate. If the shiphandler needs to hold the stern laterally in place until the entire ship is clear of the buoy line, the six shiphandling actions tell the conning officer to toe-in the waterjets to check the pivot. Once the stern is clear, toeing-out the waterjets will double the effect of pushing the bow to starboard, and the ship will snap smartly into the channel. The conning officer can check the stern's lateral movement with a quick toe-in maneuver when the ship is fair in the channel. Now that the ship is clear of the slip, both engines can be directed ahead and the bow thruster can be retracted (or the tug cast off).

Consider the same undocking evolution, but this time with an offsetting wind. It is conceivable that, with enough wind, the ship may not require any engine power to get off the pier, but this is a rare circumstance. Even if the wind is blowing nearly perpendicular to the ship, the uneven sail area of the ship will cause one end of the ship to be affected more than the other. On the *Independence* variant, the stern has a larger sail area than the bow, so a wind on the starboard side will move the stern more quickly than the bow, causing the ship to pivot to starboard.[8] Even if there is a considerable offsetting wind, it is always best to apply at least some lateral force to move the ship away from the hazard and ensure that the ship is moving away evenly, particularly given the trimaran design of the *Independence* variant.

The offsetting wind will create an additive lateral force, so the conning officer should keep in mind that any lateral controllable forces—waterjets toed-out, bow thruster trained to 270 degrees relative, or a tug pulling away to port—will result in lateral movement that is quicker than the controllable forces alone will produce. The conning officer should be ready to apply counterforce into the wind to prevent the ship from quickly crossing the center of the slip; again, the six shiphandling actions can assist in applying these forces. As the ship is being blown to port, the conning officer knows that toeing-in the waterjets will stop the stern from moving to port and that directing the bow thruster to 090 degrees relative (or directing the tug to push toward to starboard) will check the movement of the bow. The six shiphandling actions constitute a reliable method for controlling the ship in any weather conditions.

Now, consider this same undocking evolution with an onsetting wind. This condition, of course, is the most challenging, because the wind will continually push the ship toward the pier hazard. Initially, it may feel as if the ship is having difficulty breaking free of the pier. While this ship class is arguably the most maneuverable ship that the Navy has ever built, it is not immune to limitations imposed by uncontrollable forces, such as an onsetting wind too great to overcome.[9] Many factors will determine the ship's ability to get under way on its own power against an onsetting wind, but as a rule of thumb, the ship can walk laterally against a wind of up to ten knots without a thruster. With the thruster, the *Independence* variant can walk laterally against a twenty-knot wind. Wind exceeding these parameters will likely require the use of tugs to maintain positive control over the ship.

The example above describes undocking as a fairly constrained maneuver, where the ship walks laterally to port until centered in the slip and then begins moving forward into the channel, effectively driving in right angles. Thinking through the maneuver in these terms helps the novice shiphandler understand how to use the six shiphandling actions individually to control the ship, and as will be explained in Chapter Nine, on simulator training, this approach is a great way to practice precise shiphandling. The advanced shiphandler knows, however, that the practical aspects of undocking the ship include the need to get under way, move into the channel, and exit port expeditiously. Among other things, harbor pilots have other ships on their schedule, and holding them on board longer than necessary does not foster a good relationship, particularly with one who works with the ship on a regular basis. While driving in squares is good for the novice shiphandler, it is not good for a pilot "on the clock."

More realistically, then, the ship will begin moving toward the channel as soon as clear of the pier hazard, as little as twenty feet away and far from the center of the slip. This maneuver will require the conning officer to crab the ship, executing two actions from the six shiphandling actions simultaneously—moving laterally to port *and* moving ahead. While the job seems complex to the novice ship-handler, the six shiphandling actions help get it done. In this case, the conning officer moves the stern laterally to port by toeing-out both engines and moves the bow laterally to port by training the thruster to 270 degrees relative or pulling away to port with the tug. Once the ship is clear of the pier hazard—by, say, twenty feet—the conning officer can build headway by increasing thrust on the starboard ahead engine. With the port engine back T1 and the starboard engine ahead T2, the ship will begin moving ahead between one and two knots at the same time that it moves laterally to port, as illustrated in figure 5-2.

DOCKING

As mentioned in the previous section, undocking is by far easier than docking. Moving the ship toward the pier is a special challenge, because it is one of the few instances when the shiphandler intentionally moves *toward* a hazard. This maneuver must be carefully controlled. Excessive lateral speed can result in damage to the ship, particularly for the *Independence* variant, with its aluminum hull. Prevailing uncontrollable forces often preclude ideal landings, but the advanced shiphandler should strive to make contact with the fenders at a lateral speed no more than one-tenth of a knot.

While docking the ship may be tense, it should never be visually exciting. The previous section discussed considerations for getting the harbor pilot expeditiously on his or her way, but the shiphandler should never endanger the ship to ensure that the pilot stays on schedule. A delayed pilot pickup on the next ship is always preferable to damage incurred on *your* ship.

With the exception of the lateral-closure hazard, the docking evolution is very similar to undocking. Let's begin by considering a ship docking starboard side toward the pier with negligible uncontrollable forces. Of course, before entering port the shiphandler should review the six shiphandling actions for walking the ship to starboard. The conning officer would aim to drive the ship parallel to the pier until lined up alongside the assigned berth. If the pier master is particularly good, the berth will be marked by something that indicates where the bridge should line up—as sophisticated as

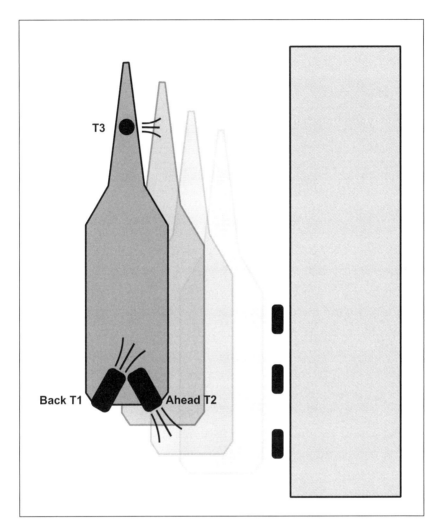

FIGURE 5-2   Crabbing Forward-Port

a "bridge here" sign or as practical as an orange traffic cone. If no marker exists, the harbor pilot should be able to identify a distinguishing feature on the pier to use as an alignment reference. The conning officer will know that the ship is longitudinally in position when this marker is adjacent to the bridge.

If the ship enters the slip using the toe-in/toe-out method to maintain precise control over the stern, with the port engine ahead and the starboard engine back, the conning officer should be using a stronger ahead engine than backing engine to maintain headway into the slip. Just before the ship is longitudinally in position, the conning officer should reduce the port ahead engine to T1 and increase the starboard backing engine, to check the ship's headway. Once the forward motion is halted, the conning officer should even out the engines to maintain a zero longitudinal resultant force vector throughout the lateral walk to starboard.

As with the undocking, the conning officer should focus on the three gauges for pierwork—heading, longitudinal speed, and lateral closure. If the pier heading is at 090 degrees true, the conning officer should aim to keep the ship's head between 085 and 095 degrees true, the longitudinal speed at zero knots, and lateral closure slowly toward the pier. This closure can be swift at first but should be slower as the ship gets closer to the pier; the ideal landing would be as soft as possible.

To begin the move to starboard, the conning officer would toe-out both waterjets with the thruster trained to 090 degrees relative (or direct the tug to push toward to starboard). If the stern begins outpacing the bow, the toe angle should be reduced; if the bow begins outpacing the stern, the thruster power should be reduced (or the tug should be stopped). With the port engine ahead T1 and the starboard engine back T1, if the ship begins sliding ahead, the starboard engine should be increased to back T2. Conversely, if the ship begins sliding astern, the port engine should be increased to ahead T2. The conning officer should aim to return to the baseline engine configuration as soon as the ship is back in position.

As previously mentioned, the danger in docking the ship is inherent in the fact that the shiphandler is moving toward the hazard. So, as the ship gets closer to the pier, the conning officer must slow the lateral movement, which can be accomplished using the by-half technique described earlier in this chapter. As the ship gets closer to the pier, the conning officer will reduce the toe angle that is laterally propelling the ship. For example, let's say the ship is laterally walking to starboard with the engines toed-out twenty degrees and the thruster ahead T4 and that the ship is moving laterally at about 0.3 knots. The shiphandler will want to slow, but not stop, the starboard movement when the ship is within thirty feet of the pier. If the waterjets are reduced by half to toed-out ten degrees and the thruster by half to ahead T2, the ship will continue moving laterally to starboard but at a slower rate—say, at 0.2 knots. When the ship is twenty feet from the pier, the conning officer can again reduce by half the toe angle to five degrees and the thruster to ahead T1, which will slow the lateral closure rate even more. When the ship is within ten feet of the pier, if the lateral closure rate has not slowed to within one-tenth of a knot, the conning officer can stop the thruster on the bow and "tap the brakes" on the stern by centering the waterjets, just in time to put over the mooring lines. If the ship is moving entirely too quickly toward the pier, the conning officer can "slam on the brakes" by toeing-in the waterjets and directing the thruster away from the pier.

Consider the same docking maneuver with an onsetting wind. The novice shiphandler may think that an onsetting wind pushing toward the berth would be preferable to an uncontrollable force fighting against the ship, but the experienced shiphandler prefers an offsetting wind any day. The offsetting wind acts as a cushion for the landing, whereas an onsetting wind can cause the ship to accelerate uncontrollably toward the pier hazard. As described above, the ship without a bow thruster can hold up laterally against ten knots of wind, and one with a thruster, against twenty knots of wind. Anything greater than twenty knots of wind will require the use of tugs to ease the ship on the pier.

Referring back to a key principle presented earlier in this chapter, the shiphandler should align the controllable forces to oppose the uncontrollable forces when docking with an onsetting wind. In this scenario, with the wind pushing the ship from port to starboard, the conning officer would position the controllable forces from starboard to port, toeing-out the waterjets with the starboard engine ahead and the port engine back and the bow thruster trained to 270 degrees relative (or tug ready to pull away to port).

Aligning the waterjets and bow thruster (or tug) in this manner will enable the controllable forces to act as a braking mechanism as the wind accelerates the ship laterally toward the pier.

This maneuver would begin with the ship in the center of the slip and aligned with the berth, the waterjets centered, and the bow thruster (or tug) stopped. The onsetting wind pushes the ship toward the pier, and the conning officer's central task is to keep this accelerating force from driving the ship into it. It is highly unlikely that the wind will blow the ship dead square toward the pier; the wind is more likely to cause the ship to pivot as it moves to starboard. Again, the three gauges for pierwork will help the conning officer to ensure the ship is walking evenly toward the pier. If the heading begins falling off to port as the ship closes the pier, the stern is being pushed more quickly than the bow, so the conning officer must toe-out the waterjets to slow the stern's lateral movement and allow the bow to catch up. Conversely, if the heading is falling off to starboard as the ship closes the pier, the bow is being pushed more quickly than the stern, so the conning officer should increase thrust on the bow thruster to 270 degrees relative (or order the tug to pull away to port) in order slow the bow's lateral movement and allow the stern to catch up.

The wind will push the ship relentlessly toward the pier, so the process for slowing the ship with an onsetting wind is the inverse of the last example. Let's say that the ship is accelerating later-ally and evenly toward the pier with the waterjets centered and the bow thruster (or tug) stopped. When the ship is thirty feet from the pier, the conning officer will toe-out five degrees and increase power on the thruster to ahead T1 at 270 degrees relative (or direct the tug away to port), so the ship will continue moving toward the pier but more slowly. When the ship is twenty feet from the pier, the conning officer would increase the toe angle to ten degrees and increase thrust on the bow to T2, which will slow the lateral closure rate even more. When the ship is ten feet from the pier, the conning officer would increase the toe angle to twenty degrees and increase the bow thruster to T4; if this is not enough power to stop the ship, the waterjets can be toed-out to thirty degrees and the bow thruster increased to T10 (or direct the tug away to port at half) to stop the ship immediately.

Now, consider the same docking evolution with an offsetting wind—which, again, is preferable. The engines would be toed-out, with the port engine ahead and the starboard engine back, and the bow thruster trained to 090 degrees relative (or tug prepared to push toward to starboard). Greater power will be required to fight the wind; while benign conditions may call for T1 on both engines, maneuvering against the wind may require T2 or even T3 on the waterjets. The conning officer may need to work the bow thruster at its maximum power (or use more than minimum tug power) if the wind is blowing hard.

Since the wind is cushioning the ship from the pier, the ship-handler can be more aggressive at the beginning of the maneuver, beginning from the center of the slip and aligned with the berth. The by-half technique is useful here as well. The conning officer would start with the waterjets toed-out thirty degrees and the bow thruster ahead T8 (or tug toward to starboard) to establish momentum toward the pier. Once the ship is moving laterally to starboard, the conning officer could reduce the waterjet angle to fifteen degrees and the bow thruster to T4, reducing it by half again if momentum continues to build. This momentum should be steadily decreased as the ship gets closer to the pier, with the ultimate goal of making contact with the fenders at a lateral closure of about a tenth of a knot. If the ship's momentum stalls at any point, increasing lateral force on the bow or stern should be applied. If the wind overcomes the controllable forces and starts blowing the ship away from the pier, the by-half technique should be employed aggressively to restore the proper closure rate: maximum toe angle and thruster power should be applied until momentum away from the pier stops, and then the forces should be reduced by half to control lateral speed toward the pier.

For all of the above examples, use of a tug is similar to that of the bow thruster only in that both apply a controllable force in the same direction; the fine control of the bow thruster is not available from a tug. This is largely due to the high thrust-to-weight differential between today's modern Z-drive tugs and this ship class. Tug management will be discussed in more detail later in this chapter, but for now it is sufficient to understand that the above maneuvers involve applying the least amount of tug force, then stopping the tug, then repeating.

Chapter Two discussed in depth the element of wind, but a note is warranted here about the nature of this uncontrollable force. Wind rarely blows at a consistent velocity; the textbook solutions presented above are intended simply for illustrative purposes. On fairly benign days, wind speed will vary by a few knots both above and below an average speed. On more violent days, the wind can gust dozens of knots above the sustained wind. This variation prevents the shiphandler from choosing one perfect combination of controllable forces that produces a perfect landing. The more the wind varies during the landing, the more the conning officer needs to adjust the engines and thruster (or tug) to maintain control of the ship. Despite this disparity between the textbook scenarios and the complexity of real-world shiphandling, these techniques will work in all conditions. For the novice shiphandler, two key methods will help dock the ship safely every time: focusing on the three gauges for pierwork and identifying the six shiphandling actions ahead of time.

A STRATEGY FOR SUCCESSFUL DOCKING AND UNDOCKING FOR THE NOVICE SHIPHANDLER: FIRST STEPS

The six shiphandling actions described above are intended to help novice shiphandlers keep the controllable forces straight in their minds. Events move quickly when docking and undocking, and the conning officer will benefit from working out the vector management ahead of time on a worksheet for reference during the actual evolution. This approach to shiphandling for the novice—simplifying the problem as much as possible so as to eliminate avoidable errors—is reason enough to reduce the theoretically infinite number of waterjet combinations to a more manageable set. This approach can be taken one step farther by consistently putting each engine in the same ahead or astern direction.

It is a common error for novice shiphandlers to confuse the lateral movement produced by a toe-in or toe-out configuration by forgetting in the heat of the moment which engine is ahead or astern and choosing the toe-angle configuration opposite to what is needed and thereby causing the ship to move laterally in the wrong direction. With the pier hazard close aboard, an error like this could be expensive. This mistake can be eliminated by consistently aligning the engines such that a toe-out configuration will always make the stern laterally move in the same direction, and the same for a toe-in alignment.

To achieve the maximum simplicity, the conning officer can arrange the engines such that *toeing-in* will always make the stern *move in* toward the pier and *toeing-out* will always make the stern *move out* from the pier. For example, if the ship were mooring starboard side toward the pier, the shiphandler would begin with the starboard engine ahead and the port engine back. When ready for the stern to *move in*, the conning officer will simply *toe-in*. If the stern closes too quickly and must be *moved out*, the conning officer simply *toes-out*. The advanced shiphandler may want more flexibility given certain environmental conditions, but this simplicity empowers the novice shiphandler with an unforgettable rule of thumb: *toe-in to go in, toe-out to go out*.

LINEHANDLING

The maneuverability of this ship class can make linehandling pretty simple. The ship's ability to walk laterally to port or starboard means that it can position itself adjacent to the pier and hover in place, even with an onsetting wind, close enough to pass lines over without committing itself to the landing. This flexibility gives the linehandling team plenty of time to get their lines over safely.

To understand this final piece in docking the ship, consider a maneuver where the ship is laterally walking to starboard against a ten-knot offsetting wind, with the engines toed-out, port ahead and starboard back, and the thruster trained to 090 degrees relative (or tug pushing toward to starboard). The conning officer will need to create a resultant force vector to starboard that is just slightly stronger than the offsetting wind to produce a closure rate around 0.3 knots. Once the ship is twenty feet off the pier, the conning officer will reduce the toe angle and power on the thruster (or tug) to produce a resultant force vector that is equal to the offsetting wind, such that the closure rate is zero. With the ship holding in this position, the officer of the deck will direct the linehandlers to pass the forward and aft breast lines and heave around. The conning officer can carefully increase the lateral force toward the pier to ease the strain on the lines by slightly increasing the toe angle and power on the thruster or tug.

When the ship is on the fenders, the conning officer may need to move slightly forward or aft to optimize the linehandling configuration to align the ship to place the brow and, more importantly in the case of the *Independence* variant, lower the side door. The conning officer can move the ship along the pier by, as indicated in the six shiphandling actions, increasing thrust on the ahead waterjet to move forward and on the astern waterjet to move aft. The bow thruster on the *Independence* variant can provide even more precise control for final adjustments, trained to 000 degrees relative to move ahead and 180 degrees relative to move astern. The thruster, though less powerful than main propulsion, allows the shiphandler to position the ship precisely.

Once the ship is in position, the conning officer should toe-out thirty degrees and train the thruster to 090 degrees relative (or tug pushing toward to starboard) to hold it tight against the fenders. Using power to keep the ship in position will allow the linehandling team to pass the spring lines and then take slack out of all lines. This function can also be performed by using tugs on the bow and stern to push the ship up against the fenders. Given their impressive thrust-to-weight advantage, tugs become a more attractive option as the offsetting wind gets stronger.

Linehandling for undocking this ship class—a shiphandling evolution that is already straightforward—is even simpler. When the ship is ready to get under way, the prevailing winds will dictate how the shiphandler should use the ship's propulsion or tugs to assist the linehandling team. With benign conditions or an onsetting wind, the linehandling team should be able to remove the lines fairly easily, but with an offsetting wind the lines may be too taut to pull them off the bollards safely. The shiphandler may need to use the waterjets, thruster, or tugs to hold the ship against the pier while the lines are removed, just as described above for making the lines fast. The officer of the deck can direct the linehandling team to take in all lines when the line tension is safe; once the lines are clear, the conning officer should proceed expeditiously with laterally walking away from the pier.

Although the maneuverability of this ship class obviates the need for complex linehandling, mooring lines represent dynamic controllable forces available to the shiphandler. They can, in fact, be very helpful in controlling the ship when alongside the pier. Lines can be employed to keep the ship from moving toward danger, and they can also be used to move the ship into a desired position. Employing mooring lines well is a skill that has become less common over the years, with shiphandlers preferring simply to take in all lines when undocking and put over all lines when docking. Dismissing mooring lines as a shiphandling tool, however, eliminates one of only five controllable forces available.

One of the simplest maneuvers employs a spring line to hold the ship in place. Refer back to figure 2-2 in Chapter Two, on controllable and uncontrollable forces, and recall that line two on the *Freedom* variant extends from the bow aft to the pier. This spring line is used to prevent the ship from moving forward while moored to the pier, but the shiphandler can use it to hold the ship from surging ahead while the stern is moving laterally away from the pier. This tool is useful for propeller-driven ships, which do not have the ability to walk laterally, and for any ship when there is a pier hazard ahead.

For example, if the ship is moored very close to the foot of the pier with the quay wall immediately in front of it, as illustrated in figure 5-3, the shiphandler can take in all lines except line two and use it as a safety net to prevent the ship from surging forward into the quay wall. Since spring lines lie at an angle to the pier, they will swing out farther from the pier when perpendicular, so the ship

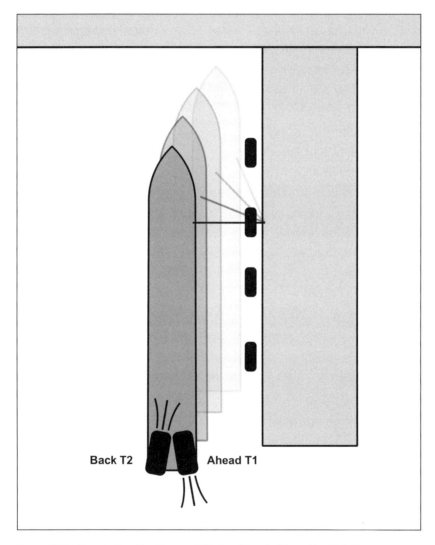

Figure 5-3   Employing the Forward Spring Line to Keep the Ship from Surging Ahead during Undocking

can walk off the pier and slide away from the quay wall—crabbing aft and to port—while line two is still singled up. The shiphandler can take in line two once the ship is well clear of the quay wall.

Consider also how the spring line can be used as a controllable force that can move the ship into an intended position when docking. Let's say that *Independence* is docking in the berth described above, starboard side toward the pier with the quay wall immediately ahead, as illustrated in figure 5-4; the bow thruster is unavailable. The conning officer would approach the pier aiming the bow at a point opposite the bollard where line four will make up. Now, the length of the line is important in this maneuver, and doing the math beforehand will ensure a safe landing. For this example, line four has already been made fast to the ship's bitts, with a length from the bitts to the eye at its end that is equal to the expected distance from the bollard to the ship's bitts when on the fenders. When the ship is in position, the line is passed to the pier and the eye placed on the bollard. As the ship continues to move ahead, the line will pull the bow snug to the pier, and the conning officer simply needs to move the stern laterally toward the pier to complete the landing. The added bonus in this maneuver is that the spring line acts as a safety net to prevent the ship from surging into the quay wall ahead.

As evidenced in these examples, mooring lines are useful controllable forces that can both prevent the ship for moving into danger and actively maneuver the ship into position. Using lines, however, adds complexity to the challenge of managing waterjets, thrusters, and tugs. As with communicating with a tug, the conning officer must account for the delay involved in sending orders down to linehandling stations. Tug captains and linehandlers diverge, however, in sustained proficiency. The shiphandler may not employ a tug during a docking or undocking, but that tug captain may work another half-dozen ships that day, one of which will likely actually use his or her services. In contrast, every day the ship does not exercise the linehandlers in more than "Take in all lines" or "Over all lines," the linehandling teams will lose more proficiency. Active linehandling takes practice, both at the

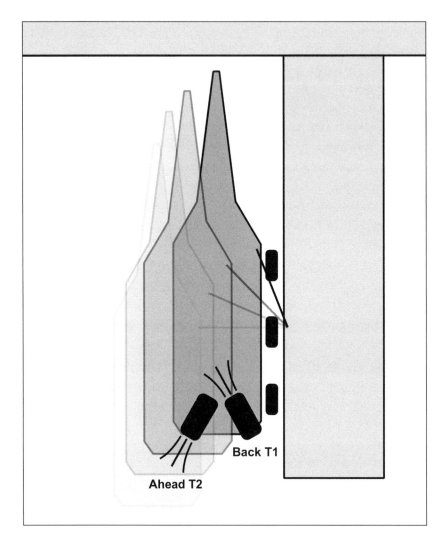

Figure 5-4   Employing the Spring Line to Pull the Bow in during Docking

linehandling stations and in communicating with the bridge. If ignored, these skills will atrophy and be absent when needed most.

TUGS AND PILOTS

As discussed in Chapter Two, using tugs during pierwork brings both benefits and hazards. As a benefit, tugs provide instant and significant power that can rescue the ship from danger nearby. Two tugs pulling on the ship simultaneously at their maximum power can exert far more lateral force than the ship's own controllable forces alone. Tugs are particularly useful if the ship experiences an engineering casualty and the shiphandler simply needs to pull the ship away from the pier hazard in order to investigate the casualty and determine its impact on the ship's controllable forces. Many experienced commanding officers will attest that tugs have saved the day when an allision seemed imminent.

As a hazard, Z-drive tugs can damage the lightweight littoral combat ship if they apply too much power for too long. Tugs pulling or pushing even at their lowest power can cause the ship to pivot more quickly than the shiphandler expects, even shifting the pivot point longitudinally. The pivot point may shift so quickly that it is nearly impossible to leave tug power applied to the ship for longer than about ten seconds.

Consider what this effect might look like during a docking evolution. Let's say that the ship is laterally walking to starboard toward the pier in a toed-in configuration, with the starboard engine ahead and port engine back. If the bow begins moving more quickly toward the pier than the stern, the conning officer will call for the tug to pull away to port. The power of the tug, if left on too long, will transition the ship's motion from a walk to a twist. In this twist, as the bow moves to port, the stern will move to starboard. The stern will now close the pier very quickly, because the lateral force vectors on the stern have become additive; twisting the ship with the tug created a lateral force on the stern to starboard when the toed-in configuration was already producing a lateral force to starboard. These combined force vectors will accelerate the stern toward the pier. In this example one can see how the tug complicates shiphandling—saving the bow endangered the stern.

For the novice shiphandler, this consequence can be both surprising and confusing. The experienced shiphandler, however, will anticipate it and take preemptive action, applying tug power on the bow and then stopping the lateral force to starboard on the stern. Using the example above, the conning officer would order the tug to pull away to port and then immediately toe-out five degrees, creating a slight lateral force on the stern to port. While that force may not precisely match the force created by the tug on the bow, at least the shiphandler has prevented additive force vectors on the stern.

The potential hazard of using tugs is an even greater consideration when operating in remote regions of the world. Language barriers often obstruct communications between the shiphandler and the tug captains, particularly when working through an intermediary, such as a pilot. Given how quickly assistance can turn into hazard, the ability to control tugs during pierwork must be considered. If the tugs are only employed as a safety net—not actively used to handle the bow or stern individually but simply tied on to rescue the ship if the docking or undocking goes awry—it may be best to have them stand off. Otherwise, the tug may be the reason that the maneuver went awry.

The power of these tugs should also be weighed when deciding how many tugs to use and where to place them relative to the ship. One modern Z-drive tug is sufficient to pull the entire ship laterally, so it is possible to use only one tug when employing it as a safety net. Consider a situation where the ship is laterally walking to starboard under its own controllable forces and suddenly loses all power. The shiphandler would need the tug to pull away to port. If the tug were tied forward of the pivot point, the stern would pivot into the pier; if it were tied aft of the pivot point, the bow would pivot in. With the tug made up amidships, as close to the ship's natural pivot point as possible, the lateral force applied by the tug would barely move the pivot point, and the ship would pull straight off the dock.

The potential hazards listed here could lead the novice shiphandler to conclude that tugs are more trouble than they are worth, but tugs are a very effective controllable force and should not be avoided simply because of what might go wrong. Instead, the shiphandler should learn to temper the power of the modern Z-drive tug. Two techniques are critical when employing tugs. First, unless the ship is in imminent danger, use the least power available—dead slow. The shiphandler can always dial up the power if the ship is in danger. Second, do not wait to see the effect of the tug on the ship before removing its force. Anticipating shiphandling effects is a skill that takes years to develop, but once the shiphandler understands when a given force will take effect and how long it will take for the order to get from the conning officer's mouth to the tug captain's throttle, it will be clear that the order to stop the tug must come long before the ship's momentum noticeably changes. Practicing over time, the advanced shiphandler can expertly control this overwhelming force to maneuver the ship.

MOVING THE PIVOT POINT

Chapter Three, on waterjet vector management, discussed how to move the pivot point on demand, and in this discussion on pierwork it is worth considering how the conning officer can affect its movement. The shiphandler must always know where the pivot point is in order to anticipate how the ship will react to uncontrollable forces; shiphandlers are even more empowered when they can choose where to place the pivot point and apply forces to move it there.

As in most shiphandling evolutions, the conning officer should begin with the six shiphandling actions for the given maneuver. Let's say that the ship is moored starboard side to a quay wall at the foot of a slip between two piers. Before beginning the undocking, the conning officer should prepare a list of six shiphandling actions similar to table 5-2 above, given that the starboard engine will be ahead and the port engine back in this example.

Moving the pivot point to the bow effectively means making the stern perform all the movement and keeping the bow steady, as depicted in figure 5-5. To move the stern to port, the six shiphandling actions guide the conning officer to toe-out the waterjets. Moving the stern to port will cause the ship to twist about a pivot point, and the bow will move starboard. The six shiphandling actions also tell the conning officer that training the bow thruster to 270 degrees relative (or directing the tug to pull away to port) will move the bow to port, but applying just enough power to

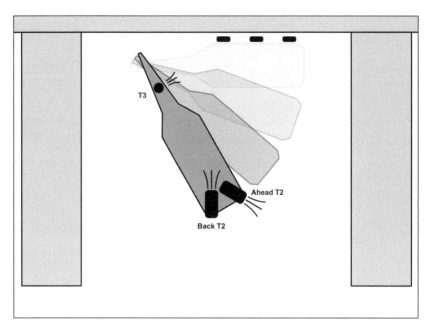

FIGURE 5-5   Moving the Pivot Point to the Bow

check the starboard bow movement will hold it in place as the stern continues moving to port. Using the three gauges for pierwork, the shiphandler will know that the pivot point is on the bow when (1) the heading is increasing, (2) longitudinal speed is zero, and (3) lateral separation is constant on the bow and opening on the stern.

Consider the same undocking maneuver, but this time the shiphandler will move the pivot point to the stern, as illustrated in figure 5-6. By definition, the pivot point will be on the stern when the bow is moving to port, and the stern will remain in place. The conning officer will begin by training the thruster to 270 degrees relative (or directing the tug to pull away to port). This force will cause the ship to twist about a pivot point, and the stern will move to starboard toward the quay wall. The six shiphandling actions tell the conning officer that toeing-out both waterjets will move the stern to port, so in order to check the stern movement, the waterjets should be toed-out just enough to counteract stern closure toward the quay wall. Using the three gauges for pierwork, the

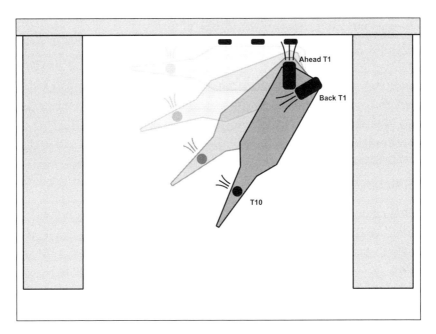

FIGURE 5-6   Moving the Pivot Point to the Stern

conning officer will know that the pivot point is on the stern when (1) the heading is decreasing, (2) longitudinal speed is zero, and (3) lateral separation is opening at the bow and constant at the stern.

Moving the pivot point during docking follows the same principles but in reverse. For simplicity, consider a docking evolution where the engine configuration is the same—starboard engine ahead and port engine back—and the ship is returning to the same quay wall berth. This configuration would still employ the six shiphandling actions outlined in table 5-2. Driving into the slip, the conning officer would aim the bow to the position just off the quay wall where line one will be made fast to the bollard. The shiphandler would stop the forward movement of the ship and move the pivot point up to the bow, which by definition means that the stern is performing all the movement as the bow remains in place. To effect this movement, the conning officer would toe-in both engines to move the stern to starboard, which will cause the ship to twist on the pivot point and push the bow to port. The six shiphandling actions guide the conning officer to train the thruster to 090 degrees relative (or direct the tug to push toward to starboard) and apply just enough power to check the bow's swing. Using the three gauges for pierwork, the conning officer will know that the pivot point is on the bow when (1) the heading is decreasing, (2) longitudinal speed is zero, and (3) lateral separation is constant at the bow and closing at the stern.[10]

DYNAMIC BOW-THRUSTER MANAGEMENT

This discussion has largely constrained bow-thruster employment to either port or starboard, at 270 or 090 degrees relative. For the novice shiphandler, this constraint provides a certain simplicity that helps develop an understanding of waterjet vector management. Once the shiphandler becomes more comfortable working with the bow thruster, however, employing it beyond the constraints of port and starboard will unlock the incredible maneuverability of the *Independence* variant.[11]

There are several factors to weigh in thruster management. For example, one of the key principles of pierwork presented at the beginning of this chapter was the importance of using the controllable forces to oppose the effects of uncontrollable forces when docking. In the case of the thruster, it is generally prudent to point the thruster into the wind. If the ship is docking starboard side to the pier with an onsetting wind from 300 degrees relative, the conning officer could train the thruster to 300 degrees relative to hold the bow up into the wind. This is particularly useful with a shifting wind; the conning officer can adjust the thruster as the wind moves relative to the ship. Since in this case the wind is what will push the bow into the pier, the simplicity of this technique lies in the fact that pointing the thruster directly into the wind means that 100 percent of the thruster is dedicated to keeping the bow from blowing downwind.

When undocking the ship, another perspective to consider is maneuvering the thruster relative to the pier hazard. The goal during undocking is to move the ship off the pier and keep it clear of hazards while it transits into the channel, so it makes sense to keep the bow thruster pointed directly away from the pier. To appreciate how the thruster can be managed dynamically while undocking, again consider leaving a quay wall and backing out of the slip, as illustrated in figure 5-7. When the conning officer begins moving the stern while holding the bow in place with the thruster trained to 270 degrees relative, the relative bearing of the quay wall will shift steadily, beginning at 090 degrees relative and ending at 000 as the ship becomes perpendicular to the quay wall and is backing out of the slip. The conning officer should swing the thruster from 270 degrees relative to 180 degrees relative at the same rate as the rate of turn. For example, when the ship's heading has moved to the right by ten degrees, the conning officer should alter the thruster by ten degrees to 260 degrees relative; when the ship's heading is off the quay wall heading by twenty degrees, the shiphandler should train the thruster to 250 degrees relative. The conning officer would follow this incremental approach until the thruster is trained to 180 degrees relative as the ship is backing out.

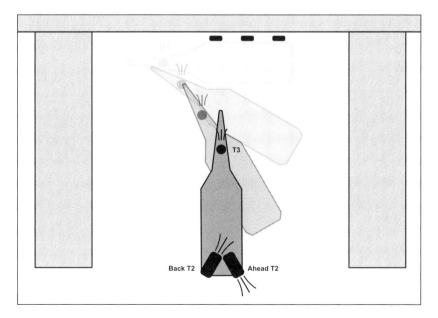

FIGURE 5-7   Dynamic Bow-Thruster Management

The added benefit of this technique is that the thruster is continually available as the ship backs out. If the bow begins falling off
to port or starboard, the thruster can be trained just to the left or
right of centerline to check the movement. For example, wind off
the port side pushing the bow to starboard will be offset by simply
training the thruster to 200 degrees relative. If the bow begins
moving too quickly to port, the by-half technique can be used
to reduce the thruster angle to 190 degrees relative, and reduced
again if necessary.

One point about vector management should be mentioned in
connection with using the bow thruster for directions other than
270 and 090 degrees relative. When the thruster is pointed either
port or starboard, the force vector contributes only to the lateral
movement of the ship. Training the thruster to some angle other
than 270 and 090 degrees relative will produce a longitudinal force
vector. Minimizing power on the thruster should provide sufficient control over the bow without considerably adding to the
ship's overall speed, but even if a noticeable force is generated, as
discussed in Chapter Two, there is little value in trying to calculate
the precise forward or aft force vector. The practical approach is
to continue using the waterjets to control longitudinal speed. If a
thruster trained to 180 degrees relative creates excessive sternway,
simply increasing power on the ahead waterjet or easing thrust on
the backing waterjet will resolve the issue. This approach does not
mean, however, that the shiphandler is relegated to being solely
reactive; the conning officer should regularly anticipate the effects
of dynamically managing the thruster. If the thruster is trained
forward of the beam, it should be expected to contribute to headway; if it is trained aft of these points, it should be expected to
contribute to sternway.

CASUALTY CONFIGURATIONS

Engineering casualties at sea sometimes require shiphandlers to
dock using less than the full complement of controllable forces.
This ship class is built with engineering plant redundancies, so

the ship can return to port safely even after the loss of multiple engines. But the nature of waterjet shiphandling means that different casualties will have different effects on what tools are available to dock the ship. Consequently, the conning officer must carefully weigh the casualty to determine how to approach the pier.

The engineering plant differences between the *Freedom* and *Independence* variants play important roles, because a casualty on one variant may be more impactful than the same casualty on the other variant. For example, the drive trains on the *Freedom* variant join the gas turbine and diesel engines on each side at a combining gear to deliver power to both waterjets on that side. This configuration allows the *Freedom* variant to suffer a casualty to either the gas turbine or diesel engine on one side and still have the outboard steerable waterjet available. On the *Independence* variant, the drive trains are dedicated from each engine to a waterjet, so losing one engine will result in the loss of its associated waterjet. In another example, because all four waterjets on the *Independence* variant are steerable, if the ship has a casualty to one waterjet, there is another steerable waterjet available on that side. The *Freedom* variant, however, only has two steerable waterjets, so losing a steerable waterjet on one side means that the shiphandler will only have one steerable waterjet available to dock the ship. The shiphandler must understand the limitations that each type of casualty creates, particularly in that the toe-in/toe-out method depends on having two engines with steerable waterjets on each side.

First, consider how the shiphandler would dock a *Freedom*-variant ship with one steerable and one nonsteerable waterjet. Recalling the discussion in Chapter Three, the conning officer can split the responsibilities for lateral and longitudinal force vectors between two waterjets. If the ship is mooring starboard side toward the pier and making its approach with the port steerable waterjet and starboard nonsteerable waterjet, the conning officer can employ the port waterjet ahead to control lateral movement and the starboard waterjet back to control longitudinal movement. To walk the ship laterally to starboard, the port ahead waterjet would be toed-out to generate a lateral force vector to starboard, which also produces a longitudinal force vector ahead. The starboard backing waterjet would cancel the forward motion, yielding a resultant force vector that is simply to starboard. Most critically, since the waterjets are not moving together as in the toe-in/toe-out method, altering the angle of the port waterjet to control lateral speed also has an effect on longitudinal speed. For example, maneuvering the waterjet from toed-out thirty degrees to fifteen degrees will reduce the lateral speed toward the pier but increase the longitudinal speed. Consequently, the conning officer must increase power on the starboard backing engine to cancel that forward movement. Should the conning officer need to increase the angle of the port ahead waterjet, decreases in power on the starboard backing waterjet will be required.

This approach is also useful when the ship has only one gas turbine and one diesel available, even when both waterjets are steerable. The challenge with using dissimilar engine types is that they produce different levels of power. The diesel provides low power at high efficiency; the gas turbine provides high power at low efficiency.[12] Employing dissimilar engines with the toe-in/toe-out method can be frustrating, because the disparity in power output makes it difficult to achieve equally opposing longitudinal forces with the two engines, which is imperative to canceling longitudinal movement when laterally walking the ship. Simply put, a throttle setting of T2 on a gas turbine is more powerful than T2 on a diesel. While either engine can be used to perform either function, the diesel engine, because it is more responsive to speed changes, is better suited to control longitudinal speed. When the ship only has one gas turbine and one diesel available for pierwork, the gas turbine is best used to control lateral movement through waterjet angle changes and the diesel waterjet to control longitudinal speed.

Finally, consider a scenario where only one engine is available to dock the ship. For the *Freedom* variant, unless the ship encountered very favorable winds, tugs would likely be required to moor safely; the tugs would control lateral movement, and the single waterjet would control longitudinal movement. The bow thruster of the *Independence* variant, however, provides some added flexibility in docking the ship without tugs. As with the toe-in/toe-out method, the bow thruster and waterjet are set against each other to cancel longitudinal movement. For example, consider a landing where the ship is mooring starboard side toward the pier using only the port diesel and bow thruster, as illustrated in figure 5-8. The port diesel would be toed-out with thrust ahead, which would provide a lateral force on the stern to starboard and a longitudinal force ahead. The bow thruster would be trained toward the starboard quarter, which would provide a lateral force on the bow to

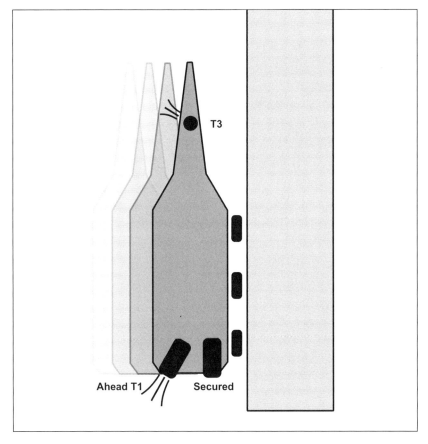

FIGURE 5-8   Employing the Bow Thruster and a Single Waterjet to Walk the
Ship Laterally

starboard and a longitudinal force astern. The waterjet would be adjusted to cancel out the longitudinal movement, thus leaving a resultant force vector to starboard.[13]

Again, the three gauges for pierwork are crucial in monitoring the ship's lateral walk, and the shiphandler must anticipate the effects of any changes made to the waterjet or thruster. If the lateral closure rate on the bow is faster than at the stern and the heading is increasing, the bow is outpacing the stern, and the conning officer must decrease the lateral force on the bow by altering the thruster toward the stern. Since this change will increase the longitudinal force vector astern as it decreases the lateral force vector to starboard, the conning officer should expect the ship to build sternway and must increase power on the port ahead engine to check that movement. Conversely, if the stern had a lateral closure rate to the pier greater than the bow and the heading was decreasing, the stern would be outpacing the bow, and the conning officer would need to decrease the waterjet angle. The adjustment will increase the longitudinal force vector ahead as it decreases the lateral force vector to starboard, so the conning officer should expect the ship to build headway and must decrease power on the port ahead engine to cancel this effect. In most cases, it will be most effective to leave the bow thruster at the same thrust setting and use the waterjet to control longitudinal speed, because the more powerful waterjet can easily check motion created by the bow thruster. The bow thruster would be hard pressed, even at higher throttle settings, to check the longitudinal force produced by a waterjet at its lower throttle settings.

CONCLUSION

This chapter began with pierwork fundamentals, building on the concepts presented in Chapter Three. With tools such as the *three gauges for pierwork*, the *toe-in/toe-out method*, and the *by-half technique*, the shiphandler can approach the pressure-filled environment of pierwork with the confidence that these fundamentals will facilitate maneuvering in any environmental condition.

The subsequent sections walked through the tasks of docking and undocking the ship, with focused discussions of linehandling, tugs, and pilots.

The true capabilities in these highly maneuverable ships emerge in the sections on moving the pivot point and dynamic thruster management, where it becomes clear how capable these ships are in austere ports with uncertain pilotage services. This ship class provides the ability to maneuver independently for pierwork and to dock in shallow ports that would be out of reach for larger surface combatants. The shiphandler who can confidently manage waterjets will be able to take these ships practically anywhere with sufficient water depth.

CHAPTER SIX

Channel Driving

O nce the ship clears the slip and enters the channel, the most dangerous shiphandling task of leaving port is complete. Driving the ship down the channel and into the open ocean is fairly straightforward, yet there are still some hazards along the way. One advantage offered by this ship class is its shallow draft, which allows it to maneuver much more flexibly outside the channel than deeper small combatants, not to mention deep-draft ships. Prudent seamanship dictates that the ship remain inside the channel, but when circumstances present more hazards inside the channel than outside, this ship class's shallow draft opens up options to sidestep those hazards.

As discussed in Chapter One, the advanced shiphandler is perpetually balancing the art and science of shiphandling, and channel driving is no different. The science of channel driving involves precise navigation tools to plot the ship's location, tools that allow the shiphandler to alter course deliberately to return the ship to the planned track. The art of channel driving is simply looking out the window and viewing the channel as a road, its edges marked by buoys. Keeping the ship in the middle of the road is as straightforward as steering between the buoys. The balance between the two approaches emerges from appreciating that while the art and science of channel driving are each often individually sufficient to exit or enter port, the advanced shiphandler can combine the advantages of both to achieve better control over the ship in restricted waters.

This chapter will explore the art of channel driving with, first, a discussion of driving by seaman's eye, a technique that requires a keen awareness of the ship's pivot point and careful attention to lateral positioning in the channel. The discussion will then turn to the science of channel driving, looking at the ship's electronic navigation and track-control systems. Finally, this chapter will examine emergencies in the channel before concluding with a special consideration of squatting.

DRIVING BY SEAMAN'S EYE

Relying on the art of ship driving is commonly referred to as "driving by seaman's eye," where the shiphandler visually processes such factors as heading, relative motion, navigation aids, and observable environmental factors to determine whether the ship is following a safe track. For channel driving, the most important element in visually determining the ship's position is the buoy line.

In a straight leg in a channel lined by buoys, the buoy line provides a range of indications that the shiphandler can use to assess the ship's progress. Consider this range of indications from end to end, with the ship outbound from a U.S. port, as illustrated in figure 6-1. If the red buoys on the left all line up and the green buoys on the right spread out (position A), the ship is on the far left side of the channel. If the red and green buoys spread evenly, with the apparent separation equal on both sides (position B), the ship is near the middle of the channel. If the red buoys spread out on the left and the green buoys line up on the right (position C), the ship is on the far right side of the channel. Between these three are many points in the channel, which constitute the range of indications from the left to the right side. Driving by the seaman's eye, the shiphandler can assess the ship's lateral position in the channel and adjust course to regain the track if necessary. Additionally, through the lens of relative motion, the conning officer will know that the ship is squarely between two approaching buoys when the bearing drift of a red buoy is equal to that of the opposite green buoy.

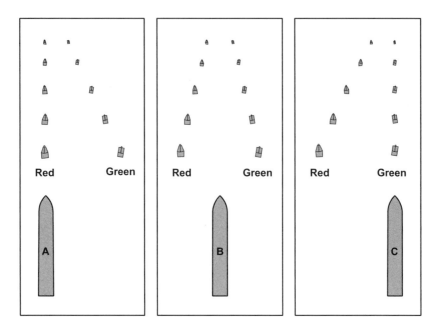

Figure 6-1   Driving the Channel by Seaman's Eye

Watching the buoy line can also help the conning officer make turns on a new leg of the channel that is marked by two or more buoys in a straight line. For example, again exiting a U.S. port, let's say that the ship is turning left onto the next leg. As the ship approaches the turn, the conning officer will see the red buoys of the next leg line up as the ship enters the channel line; if the ship does not turn on time and instead drives through the turn, the green buoys will line up as the ship exits the other side of the channel. While the ship's speed and rudder angle are important factors for timing the turn, the point when the buoys line up on the near side marks approximately when to commence the turn when driving by the seaman's eye.

One of the most reliable and accurate tools for driving by the seaman's eye is a navigational range. Generally found on long legs of major port entrances, ranges are marked by two large signs, one above and behind the other, often lighted for night transits. The marker that is above and behind serves as the ship's position indicator. If the higher marker is seen to be to the left of the lower marker, the ship is on the left side of the channel. If the higher marker is to the right of the lower marker, the ship is on the right side of the channel. If the two markers are aligned, of course, the ship is in the center of the channel. Driving by the seaman's eye often means a certain amount of error in comparison to more scientific methods, but the navigational range is the exception to this generalization. Navigational ranges are precisely placed, so they precisely indicate the ship's position in the channel.

The nature of this precision, however, should be carefully understood. Since these ranges are based on the line of sight, the position of the shiphandler evaluating the range matters, particularly for the *Independence* variant, with its wide beam. An individual viewing the range from the ship's centerline will see a different alignment than someone on the bridge wing. Since the *Independence* variant has a beam of 104 feet, a person on the starboard bridge wing may see the ship centered in the channel at the same time as an individual at centerline would evaluate the ship as seventeen yards left of track and one on the port wing would consider the ship thirty-four yards left of track. Referred to as *parallax*—apparent movement caused by changed location of the observer—this phenomenon is often used to explain why two individuals on the same ship can reach differing conclusions from accurate and objective range markers. The more practical application, however, is in understanding that no one observer's assessment about the ship's position accounts for every part of the ship traveling down the channel. If the conning officer standing on the centerline is driving outbound in the center of the channel and arranges a port-to-port passage with an inbound vessel, the shiphandling team must understand that the port side of the ship is encroaching more than fifty feet into the inbound side of the channel.

PIVOT POINT AND LATERAL POSITIONING IN THE CHANNEL

This last point highlights the fact that the shiphandler must always be aware of the ship's lateral position in the channel, from the bow all the way to the stern. The discussion of pivot-point behavior in Chapter Two, on controllable and uncontrollable forces, explained how the pivot point moves in the same direction as the longitudinal movement of the ship, so driving ahead will push the pivot point forward, until it rests about one-third the ship's length from the bow. As a rule of thumb, for every one degree that the ship alters course, the stern will swing about ten feet.

This factor is principally important in two scenarios. First, whenever the ship turns onto subsequent legs during a channel transit, the conning officer must remain aware of the stern's movement. Novice shiphandlers can become very focused on the ship's track ahead, ensuring throughout the turn that the path is clear of danger. This skill, likely ingrained when learning to drive a car, is certainly useful to keep from driving into a road hazard, but for a vehicle, as long as road traction is not compromised, the rear end will follow the front. For a ship, however, as the rule of thumb above tells us, the stern will not do the same. For example, if the ship turns forty-five degrees to the next course, the stern will laterally move 450 feet—more than the ship's length! The experienced shiphandler knows to check the side of the ship *opposite of the turn* to ensure that the stern is clear of obstructions and can swing freely onto the new course.

In the second scenario, uncontrollable forces can force the ship to drive a course considerably left or right of the intended course just to keep on the intended track within the channel. Substantial wind or current from the starboard side, for example, will push the ship toward the left side of the channel, and the conning officer must steer right of the planned course to overcome this force, a condition referred to as crabbing down the channel. If the shiphandler standing centerline observes the ship to be in the center of the channel, the bow will be on the right side of the channel, and the stern will be on the left. Again using the rule of thumb above, steering ten degrees right of the planned course will swing the stern one hundred feet into the left side of the channel, more than the beam of the *Freedom* variant. For the *Independence* variant, the port side of the stern will be more than 150 feet left of the channel's center, as illustrated in figure 6-2. The experienced shiphandler knows to check continually the lateral position along the entire ship's length in these situations.

ELECTRONIC NAVIGATION

In the balance between the art and science of shiphandling, the scientific approach to channel driving focuses on today's modern electronic navigation equipment—in this case, the Voyage Management System (VMS). This ship class was built to maximize the system's usefulness by connecting it directly to the shiphandler; the

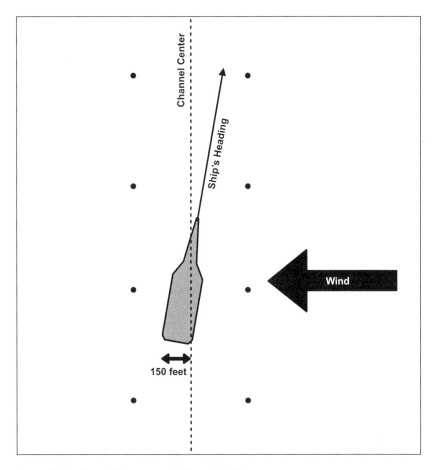

FIGURE 6-2   Stern Movement during Crabbing

displays remove the layer that traditionally lies between the con-ning officer and navigational data. Rather than positioning the nav-igation display off to the side, where a navigator and quartermaster report the ship's position, the display sits directly at the conning officer's hand.

This arrangement also eliminates unnecessary breaks between navigational reports. In a traditional watch team, although VMS updates the ship's position every second, the navigator only gives a verbal report every three minutes; the once-per-second update rate is useless to the shiphandler. On this ship class, the VMS display is directly in front of the conning officer, where it can be scanned continuously, along with other gauges. This means that the ship-handler will check the ship's position many times a minute and can be much more responsive to undesired ship movement.

Chapter Five, on pierwork, discussed how the conning offi-cer can become overwhelmed by the enormous amount of data available from ship systems and how narrowing this data down to the three gauges for pierwork helps focus the shiphandler on the information that provides the most benefit for laterally walking the ship. Similarly, the VMS display can be overwhelming, not just to the novice shiphandler but to anyone who is unaccustomed to it. When driving down the channel, the conning officer should focus on three particular data points for channel driving—cross-track distance, distance to the next turn, and set/drift.

Cross-track distance provides up-to-the-second measurements between the center of the ship and the plotted track in the channel. Like most distances observed in shiphandling, it cannot be consid-ered as a single measurement in time. Instead, it must be weighed as a series of measurements across time, because the ship is open-ing, closing, or steady relative to the plotted track.

For example, if the conning officer notices in one scan that the ship is twenty yards right of track and then a minute later that the ship is thirty yards right, the cross-track distance should be evalu-ated as thirty yards right and opening. The shiphandler knows that action is required or the ship will soon be forty yards right of track. If, however, the first scan shows that the ship is twenty yards right of track and the next ten yards right, the ship is ten yards right of track and closing. In this case, the shiphandler knows that the ordered course is moving the ship back toward the plotted track but also must anticipate that holding this course much longer will place the ship on the left side of the channel. In the last possible situation, where the ship is holding steady relative to the track, the conning officer knows that the ordered course is neither improving nor worsening the ship's position, and the actual distance from the track will determine whether this is acceptable.

For the novice shiphandler who lacks perspective as to cross-track distance, approaching traffic and the width of the channel will determine what the acceptable margins are. As a rule of thumb, however, cross-track distance less than ten yards is effectively *on track*. Distances between ten and twenty yards are *slightly left/right of track* and can be corrected without taking substantial action. Distances greater than twenty yards should be reported to the entire shiphandling team, and the conning officer should maneuver the ship to regain track.

Distance to the next turn provides the shiphandling team situational awareness as to when the ship must make its next major maneuver. This distance is measured from the ship's position to the point where the conning officer must turn the waterjets to set the ship on the next leg.

Distance to the next turn informs several decision-making factors in channel driving. First, it will determine how much time the ship has to correct for cross-track distance prior to making the next turn. If a ship traveling at ten knots is off track by twenty yards with three thousand yards remaining in the leg, then it has nine minutes to regain track, so only a small course correction is needed. If there is only a thousand yards remaining, the ship has only three minutes to regain track, and a more substantial course correction is required.[1] If only one hundred yards remain, it is likely too late to regain track, and the shiphandler should adjust the turn point to account for being out of position. If the ship is on the opposite side of the track with respect to the next course—right of track and approaching a turn to port or left of track approaching a turn to starboard—the conning officer should turn earlier than planned. Conversely, if the ship is on the same side as the next course—left of track for a port turn or right of track for a starboard turn—the turn should be made later than planned.

Second, distance to the next turn serves as a prompt for actions required to turn the ship safely. Prior to turning more than ten degrees, the conning officer—or any member of the shiphandling team—should check the side to which the ship is turning to ensure that no overtaking vessel has slipped into a blind spot. Someone should also check the side away from the turn to ensure that there is sufficient room for stern swing.

Finally, the conning officer should weigh how the upcoming turn will impact intentions with regard to passing vessels. For example, if the ship has arranged by radio a port-to-port passage with an oncoming vessel, the conning officer should consider whether a turn to port is advisable. Even if such a turn is determined to be safe, the conning officer's intentions must be communicated to the oncoming vessel, whether by radio, sound signal, or management of the ship's heading so as always to present a safe target angle (port aspect, in this case) to the oncoming vessel.

Set and drift measure the ship's movement caused by uncontrollable forces. "Set" is indicated as degrees true in the direction that the ship is being pushed, and "drift" is the true speed, in knots, at which the ship is being pushed. These measurements do not discern what is causing the movement, which could be any combination of wind, current, or seas.

For example, if the ship is inbound in inland waters on a course of 000 degrees true and the set and drift is 090 at 0.2 knots, uncontrollable forces are pushing the ship to the right, onto the red buoys. The conning officer would need to steer left of 000 degrees true to offset this force; the ship will consequently crab down the channel. In these cases, the entire team must understand the shiphandling plan, since steering a course other than the plotted track for extended periods will seem to deviate from the briefed plan.

TRACK CONTROL

The advanced technologies of this ship class permit automated systems to assume some duties previously carried out by watch-standers. One of the more innovative features of the ship class is track control, which allows the navigation system to make heading changes automatically to keep the ship on the VMS-plotted track. This automation, however, does not relieve the shiphandling team of the responsibility to navigate the ship safely into and out of port; the shiphandler must understand how the system thinks, why it is ordering specific heading changes, and what indications point to system failure. Any computer system can fail; the watchstanders must know how to monitor its performance effectively.

Before track control can be activated, the ship must be within a specified distance of the plotted track and also within a specific number of degrees of the plotted heading for that leg.[2] Once within these parameters, the watchstander will order and acknowledge track control activation in both VMS and autopilot, after which the entire shiphandling team should be notified that the navigation system is controlling the ship's movement. The navigation system will begin making continuous heading changes to keep the ship on track.

At first these course changes, often considerable, may be dis-concerting. This system behavior is a good example of why the human on watch must understand how the system thinks. The computer measures the distance between the plotted track and the ship's common reference point, or programmed center of the ship, which is located on the ship's centerline. If the ship is right of track, it will alter course to port, and if the ship is left of track, it will alter course to starboard. The system is simply trying to keep the ship's common reference point as close to the track as possible. It orders heading changes without considering nearby buoys, recreational boats, channel shipping, or any other obstructions. Since the plotted track and ship's position are the only input factors, the shiphandler's best indicator of system performance is cross-track distance. Heading changes may vary as much as ten or fifteen degrees from the plotted course, but as long as the ship is within ten yards of the plotted track, the system is doing its job.

The predominant characteristic of the system is that it drives like a computer, not a person. In some cases, this characteristic is merely something to be understood; but in other cases it makes track control not optimal for safe channel driving. For example, it is common for track control to make course changes large enough to place buoys on the "wrong" side of the bow. The prudent ship-handler entering a U.S. port would generally point the ship such that the next red buoy remains on the starboard bow; the navigation system, as noted, does not consider buoy locations—they are irrelevant to its assigned task. As long as the plotted track remains between the buoy lines, the navigation system will keep the ship in the channel. Provided that the entire length of the ship passes safely between the buoys, meaning that course changes do not swing the stern into a buoy, the shiphandler can accept the visually disconcerting presentation for the sake of more accurate channel navigation.

Consider the effect of excessive yawing, however, when maneuvering in close quarters with channel shipping. Good mariners use target angles to communicate with each other, so the ship's head swinging back and forth may appear at least confusing, if not dangerous, to approaching traffic. For example, if two ships had arranged a port-to-port passage in the channel, each captain would expect to see the other ship's port bow. Using track control during this maneuver could be problematic, even if the track is laid on the right-hand side of the channel, because the system may unpredictably present a wide variety of target angles—port bow, then starboard, then port, then starboard—to the other ship. Although the passage had been clearly arranged, track control would thereby create uncertainty that would make the passage less safe. In these cases, the conning officer should disengage track control until after passage and resume when clear.

For those unfamiliar with automated channel navigation, relinquishing control over the ship in restricted water may seem unthinkable, but any experienced mariner knows that even the most proficient shiphandling team sometimes struggles to remain on track. Decision-making can be adversely affected by low visibility, unfamiliarity with a port, or a breakdown in bridge resource management, which can happen even with well-trained teams. As long as the track is laid precisely and the ship is receiving a good GPS signal, the system can be relied on to keep the ship within single-digit yards of the track. Delegating the duties of channel driving to track control will allow the shiphandling team to focus on other tasks, such as evaluating upcoming traffic and identifying bail-out areas in the event of an engineering casualty.[3]

EMERGENCIES IN THE CHANNEL

The discussion thus far has focused on making small course changes to keep the ship on the plotted course; like many other shiphandling evolutions, channel transits are typically straightforward, with little excitement. Also like many other shiphandling evolutions, however, these basic evolutions can be unexpectedly interrupted by emergencies that put the ship in danger. The added element during channel transits is that the ship is surrounded by shoal water, so even minor events can have a big impact. Casualties that cause a loss of ship control give the shiphandling team little time to react before the ship touches ground. Also, external hazards—weather, risk of collision with traffic in the channel, nuisance pleasure craft that either do not understand or do not respect the constraints of a ship in the channel—can force the shiphandler to maneuver in very tight quarters. These slim margins for error result in increased risk to the ship, even on the most apparently benign days.

Because of this persistent danger, the shiphandler must always know the direction to safe water. Before the advent of VMS displays, the navigator would deliver to the conning officer lengthy reports of information needed for a safe transit—the ship's location, the hazards surrounding it, and aids to navigation, among other details. The navigator also reported the bearing and range to the nearest shoal. This information is important in that it tells the conning officer where not to turn, but when an emergency develops, the more important and relevant question on the bridge is, Where is the safe water? This perspective best prepares the shiphandler to take quick action when needed.

In terms of safe water, the shiphandler should avoid focusing on the single point where a shoal is located. Instead, the conning officer should weigh the ship's complete underwater surroundings and determine a bail-out area that provides the most room to maneuver. The single-point calculation can be deceiving, for example, if there is only one limited area of shallow water near a larger area of open water, as illustrated in figure 6-3. Avoiding the nearest shoal could be as easy as clearing one more buoy, which would offer a path to a bail-out area with plenty of space to wait for a shipping hazard to clear. In the illustrated example, the option to starboard would be far superior to the opposite side of the channel,

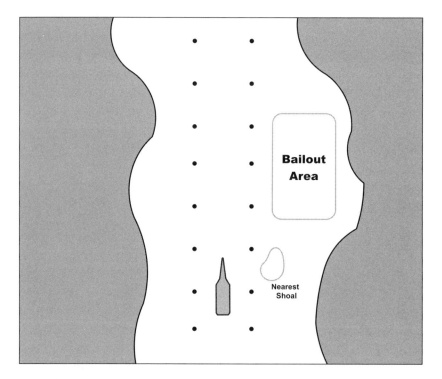

FIGURE 6-3   Identifying the Bail-Out Area

where shoal water is farther from the channel line but offers less room to maneuver.

The concept of the bail-out area is so important that it should be part of the running conversation among the shiphandling team throughout every channel transit. Each time the ship turns on to the next leg, the conning officer should look for the next possible bail-out area and discuss its location with the rest of the team. Similarly, if the bottom contour of the surrounding area changes considerably, the conning officer should update the bail-out area for the team.

To be sure, some channel transits are tight paths, with no room to either port or starboard. These channels are safe only inside the buoy line even for this ship class, with its shallow draft; that does not necessarily mean, however, that the ship is surrounded by bad water. Keep in mind that a ship on a plotted track always has good water ahead and astern, and sometimes one of those options is more palatable than the other. For example, at the beginning of a lengthy stretch of confined water, the shiphandler may determine that the bail-out area is astern. If a hundred yards into a stretch the weather makes the transit too hazardous, the smart choice may be to back out rather than drive deeper into confined waters. At some point during a constrained leg, however, the safe water might be farther behind than ahead, so the smart choice would be to press on. In any event, locating the bail-out area is important before an emergency develops.

Hovering

Another advantage of this ship class is that it does not need headway to maintain steerageway. A propeller-driven ship must, as discussed earlier, keep enough water flowing across the rudder to produce lift, which in turn controls the stern. In contrast, water-jets are always discharging water, regardless of the ship's speed, so the conning officer always has control over the stern. With water-jets opposed to cancel out longitudinal force vectors, the ship can effectively hover. The bow thruster on the *Independence* variant provides even greater control, allowing the ship to hover in more challenging environmental conditions.

This capability proves useful when bailing out of the channel toward an area that does not have much maneuvering room. For example, let's say that the ship is transiting outbound in the channel and an inbound deep-draft vessel is approaching. The ships would normally pass safely port-to-port, but if the oncoming vessel encroaches on the outbound side of the channel, the conning officer may need a bail-out option. In this case, we will say that there is just enough room to step out of the channel but not enough to haul out widely or even to drive parallel to the channel much farther. The conning officer can slide out of the channel, stop the ship, and hover until the deep-draft vessel has passed.

Hovering can be a very challenging shiphandling task. First, although this ship class maintains steerageway even when stopped, the ship is more susceptible to set and drift when dead in the water, because the conning officer instinctively uses less power. Furthermore, ships in the middle of a harbor are more exposed to wind and current than those in the protection of a slip framed by two piers. The shiphandler can overcome these issues when hovering by applying more power. Safe handling alongside the pier constrains the shiphandler to use only enough power to effect the desired movement, but without a pier hazard nearby the conning officer can confidently use greater power.

Second, the shiphandler does not have a fixed reference point nearby for assessing ship movements, which makes it difficult to determine the effects of uncontrollable forces. While they are not fixed objects, however, buoys riding on a steady strain provide some indication of the ship's relative motion, and watching the buoy line indicates when the ship crosses outside of the channel. The conning officer can also toss markers, such as wood chips or bread slices, into the water off the bridge wing to sense the ship's drift. Although these indicators are subject to current, they should allow the shiphandler to visualize the ship's movement.

Leveraging the science of shiphandling, the conning officer can also use modern tools on board to determine ship movement.

VMS provides a predicted ship display that roughly shows the direction the ship is moving and how it will move in the harbor if the present controllable and uncontrollable forces continue. Radar can also provide useful information in evaluating relative motion with respect to visible objects. For example, switching the display to relative vectors and observing the relative motion of contacts with zero speed, such as buoys, will indicate whether the ship is opening or closing those objects. In addition, most ships are outfitted with laser range finders, which the conning officer can use to measure distance to buoys and determine whether the distance to them remains steady. Finally, the shiphandler can always rely on the gyrocompass to assess drift, by shooting bearings to the buoys; left bearing drift indicates rightward ship movement, and right bearing drift indicates leftward movement. Buoy lines run nearly parallel to most channels, so if the channel runs along 180 degrees true, the buoy line should lie on a line of 180 degrees true; as the ship crosses out of the channel to starboard, the buoys will bear 180 degrees true and have left bearing drift. The buoys will bear less than 180 degrees true when the ship is outside the channel. Of course, two reliable indicators of a successful hover are a constant bearing to the buoy and no change in distance.

The three gauges for pierwork—heading, longitudinal speed, and lateral closure—are useful in assessing the ship's hover as well. Heading does not need to be nearly as precise as during pierwork, so the goal in this maneuver is to keep the ship roughly parallel to the channel heading. The more important goal is to keep both longitudinal speed and lateral closure near zero. Understanding the six shiphandling actions will help the conning officer know what controllable forces to employ to hold the ship in place. Table 6-1 lists the six shiphandling actions for hovering the *Independence* variant, with the starboard engine ahead and port engine back.

Consider how the uncontrollable forces will affect the ship's movement during a hover. Using the same scenario presented above, let's say that in wind blowing from 090 degrees true and slack current, the ship is pointed 180 degrees true, parallel to the channel, and outside the buoy line. The wind will push the ship

TABLE 6-1 Six Shiphandling Actions for Hovering the *Independence* Class, with the Starboard Engine Ahead and Port Engine Back

SIX SHIPHANDLING QUESTIONS	SIX SHIPHANDLING ACTIONS
1. What makes the stern go to port?	Toe-out waterjets
2. What makes the stern go to starboard?	Toe-in waterjets
3. What makes the bow go to port?	Thruster to 270 degrees relative
4. What makes the bow go to starboard?	Thruster to 090 degrees relative
5. What makes the ship move ahead?	Increase the starboard ahead engine
6. What makes the ship move astern?	Increase the port backing engine

to starboard and farther outside the buoy line if unchecked, so the conning officer must use the controllable forces to lean into the wind. The six shiphandling actions in table 6-1 dictate that moving the stern to port would require toeing-out both waterjets and that training the thruster to 270 degrees relative would move the bow the port.

Winds vary slightly, so the conning officer should expect to increase and decrease the waterjet toe angle to control lateral movement of the stern and to adjust thruster power to control lateral movement on the bow. For example, the conning officer may toe-out twenty degrees to stop the stern from moving to starboard, but if the engines overpower the wind and swing the stern to port, the conning officer would ease the toe angle to ten degrees. As described in Chapter Five, the by-half technique is useful in finding just the right amount of power. Similarly, setting the bow thruster at T4 might stop the bow from blowing down to starboard, but if the thruster overpowers the wind, the conning officer should ease the power to T2. If both the ahead and astern engines are T1, longitudinal speed is controlled by changing thrust on one engine or the other. If the ship slides backward, the conning officer should increase the starboard ahead engine; if the ship moves ahead, increasing the port backing engine will check that motion. Hovering in this manner without a thruster is possible, but as previously discussed, the ship will hold up laterally only against ten knots of wind or less.

Emergency Use of the Anchor

The anchor can be used for many different purposes, but it will never seem more valuable than when employed to save the ship from grounding, collision, or allision during a channel transit. Every ship in the Navy makes its anchor ready for letting go prior to entering port or undocking, but few will ever use it. This safety net is held in reserve until the other controllable forces—waterjets, bow thruster (*Independence* variant only), and tugs—have failed to regain control of the ship.

For the novice shiphandler, it may seem unthinkable that the ship would find itself in this position, with all controllable forces failing, but cascading engineering casualties can result in a complete loss of power that leaves the ship entirely subject to the uncontrollable forces. Dropping the anchor while the ship still has significant momentum can damage the anchor windlass system; this damage may be an acceptable alternative to grounding, collision, or allision, but the associated cost means that the shiphandler should refrain from this option until absolutely necessary.

Once the shiphandler makes the decision that all other controllable forces have been exhausted, the anchoring options are constrained by two factors: the ship's speed and the amount of room remaining before reaching the hazard. Attempting to set the windlass brake at greater than five knots will likely strain the anchor system beyond its structural limits; an interim option is to drag the anchor. This is accomplished by letting the anchor go and setting the brake when at short stay (see Chapter Two). The anchor will drag along the harbor floor, providing friction that slowly bleeds off momentum. Once the ship's speed falls below five knots, the shiphandler can release the brake and pay out enough chain to attempt to set the anchor. To allow the flukes to dig in, the chain must lie nearly horizontal with respect to the harbor floor, so enough chain must be paid out before setting the brake. A length of two to three times the water's depth may be attempted, but if there is enough room, the traditional rule of thumb of five to seven times the depth will provide the best chance of setting the anchor.[4]

SQUATTING

All ships experience some degree of squatting—depression of the stern—in shallow water. Squat is a function of water depth and speed and is related to Venturi forces. As discussed earlier, as the water forward of the ship is pushed out of the way, some water particles move down the side of the ship, and some move beneath the ship. They all rejoin astern of the vessel at the same location as those that were not disturbed; these particles are moving at a higher velocity. Consider again Bernoulli's principle, that in a fluid with uniform flow the sum of the following will remain constant: energy from velocity, energy from pressure, and potential energy from elevation.[5] As we concluded, since the elevation at sea level remains unchanged, this formula is narrowed to a function of velocity and pressure. The water in the ocean is subject only to two components of pressure—static pressure from the weight of the water and dynamic pressure caused by the movement of surrounding water. Changes in depth at sea are not significant enough to change the pressure of the water, so we consider the static pressure to be constant, making the changes in velocity the only influence on pressure of water in the ocean.[6] So, simply driving the ship through the water creates a low pressure area beneath the ship, and the stern drops to some degree in any depth of water. The faster the ship's speed, the more the ship will squat.

This phenomenon is compounded in shallow water, because water moving under the keel is squeezed in the narrow area between the ship and the harbor floor. As discussed in Chapter Two, water moving through confined spaces will increase in velocity. This increase in velocity will decrease the pressure under the ship even further and cause the stern to drop more than it would in deeper water. Additionally, the space beneath the keel is constrained in three dimensions, so shallow water in a narrow channel will move at an even higher velocity. That is to say, the ship will squat most at its highest speed, in water that is extremely shallow, and in very narrow channels. The shiphandler must be aware that increasing speed in shallow water and narrow channels will increase the ship's draft astern much more than if the ship were in open water.

The effects of squat are not that significant as a percentage of total draft, but they must be considered when operating the ship on the margins of navigational draft. This ship class is designed to conduct sustained operations in shallow water, so when the ship is close to the limits of its navigational draft, the conning officer must remain aware that speed will increase the depth of the stern. In applying this knowledge practically, the shiphandler must choose which factor is more important, area access or speed. If the tactical situation does not allow for constraining the ship's speed, the risk of grounding can be managed by identifying area limitations. For example, if the navigational draft is nineteen feet and the maximum change in draft at sprint speed is three feet, as indicated by the ship class squat characteristics, the navigation team should change the navigational draft in VMS to twenty-two feet.[7] The electronic chart will then indicate the area of water where the ship can navigate safely without concern that increasing speed will risk grounding. Conversely, if ensuring access to a certain area is most important, the navigation team should note how much squat in that area is acceptable. Again comparing that data to the ship class squat characteristics, the shiphandling team will know the maximum speed that would not risk grounding.

CONCLUSION

The above discussions of emergency actions in the channel center on the least frequent situations, but naval officers habitually focus on the worst cases to ensure readiness in dire circumstances. The maneuverability of this ship class provides much more flexibility in emergencies than is available to a propeller-driven ship. With the ability to bail out toward water outside the channel or even hover in place, the shiphandler can reduce relative motion between the ship and surrounding hazards, buying additional time to find a solution or allow the danger to pass.

Of all the chapters so far in this book, this chapter best embodies the balance between the art and science of shiphandling. Some shiphandlers swear by the strict science of shiphandling, relying on navigation fixes from both visual and electronic means to verify the ship's position along the plotted track most accurately. Others lean more toward the art, driving the ship down the channel as they would a car down the road, with the straightforward goal of keeping the ship between the buoys and away from other traffic in the channel. As in most things in life, the optimal solution often shifts between the two extremes based on the dominant environmental conditions of the day. Perfect visibility and absence of traffic in the channel allow the shiphandler to lean farther toward the art, but reduced visibility and dense traffic force the shiphandler to depend on the science. The shiphandler who has command of the entire spectrum is best equipped to handle any situation.

CHAPTER SEVEN

Special Shiphandling Evolutions

In the world of ship driving, there is a logical imbalance between normal operations and special shiphandling evolutions, an imbalance implied by the qualifier "special." Most days at sea are spent driving between two locations or operating in a designated area, both relatively low-risk evolutions. Mariners generally keep plenty of room between each other, and warships are notorious for increasing that buffer even more. Contacts are identified as far away as feasible, and conning officers instinctively maneuver to add space.

On the other side of this imbalance, only a fraction of an at-sea period is spent conducting special shiphandling evolutions. These events, however, require the full attention of the shiphandling team, and this focus must begin long before the event occurs. Even before arriving at the designated location, the shiphandling team briefs the event and, more importantly, the plan to execute it safely. As in the discussion in Chapter Five, on pierwork, all members of the team must know the maneuvering plan, so that they can recognize when something has gone awry.

This chapter will explore the most common special shiphandling evolutions for this ship class, the majority of it dedicated to underway replenishment. Refueling the ship alongside an oiler has the special quality of being both an inherently dangerous and commonly occurring event. Small-boat operations are also discussed in detail, focusing primarily on creating a safe lee for the boat while alongside. This chapter also examines other events—anchoring, waterborne mission-package operations, and recovery of a man overboard—that, although less dangerous and less frequent, still require the shiphandling team's close attention for safe execution.

UNDERWAY REPLENISHMENT

The ability to refuel at sea is one of the defining capabilities of a modern navy. Effective logistics are required for sustained operations at sea, and replenishing fuel and stores without entering port is critical to a force's ability to remain on station. In addition, while coast guards are designed to defend home waters, navies are built to deploy, so their ships must be able to receive fuel at sea to enable them to cover transoceanic distances.

The maneuverability of this ship class gives the shiphandler extraordinary capabilities for driving it alongside an oiler. The technological advances on this ship's bridge allow the shiphandling team to make the approach with a fraction of the watchstanders needed by a traditionally designed ship. All in all, the ship's handling characteristics provide for a swift approach, precise alongside handling, and powerful breakaway. With more power than an *Arleigh Burke*–class destroyer at one-third the weight, this ship class can execute a rapid separation from the oiler.

Scheme of Maneuver

For the novice shiphandler, it is important to understand the general scheme of maneuver for alongside replenishment. The evolution begins when the oiler assumes tactical control of the ship and orders it into standby station.[1] The ship proceeds to a point three to five hundred yards on the oiler's quarter, with a lateral separation slightly greater than the intended separation when alongside, on the side on which the refueling will occur. This ship class always takes station on the oiler's starboard quarter, since its own refueling station is on the port side. A good position would be four hundred yards on the starboard quarter, with a lateral separation of two hundred feet.

When the oiler is ready to begin and the ship is ready to take fuel, the ship will make its approach, at a speed moderately faster than the planned refueling speed. The goal in the approach is to move smartly alongside but always maintaining control of the ship. As in pierwork as described in Chapter Five, it is highly unlikely that the conning officer will find a course and speed that suffice throughout the replenishment. The dynamic uncontrollable forces at sea have a different effect on the ship than on the oiler, and minor variations in any engineering system prevent perfectly steady performance. As such, the conning officer will spend the majority of the time alongside making minor adjustments to keep the ship longitudinally aligned and at the proper lateral separation.

Once the refueling is complete and all lines have been returned to the oiler, the conning officer will commence the breakaway. The breakaway maneuver may vary based on the circumstances, but in general, the ship will accelerate quickly and turn to starboard slowly. The goal in the breakaway is to increase separation from the oiler to reduce the hazard of collision. This maneuver can be executed as rapidly as flank speed or as slowly as the approach speed, but as long as the ship is pulling ahead and hauling to starboard, the breakaway should be relatively straightforward.

This general description of the scheme of maneuver is meant simply to paint the picture of a standard alongside replenishment. The discussion below will consider the specific shiphandling methods and techniques available to the conning officer to complete the refueling safely. Before learning how to drive the ship, however, the shiphandler must understand the forces acting on it. Chapter Six, on channel driving, reviewed the practical implications of Venturi forces during channel transits, but the phenomenon is particularly dangerous between two ships in close quarters at sea.

Venturi Forces

As a ship moves through the ocean, the flow of water creates areas of increased and decreased pressure along the hull. The bow creates a high-pressure area as it pushes water to the side. This water must travel a greater distance along the outside of the hull, before returning to its original position astern of the ship, than it would have had it not been disturbed; so we know that its velocity increases. The higher-velocity water creates a low-pressure area along the hull. As a result, any ship cruising through the water will have a high-pressure area on the bow and a low-pressure area along the length of its hull.[2]

These areas of differing pressure can be hazardous when ships are operating alongside each other, such as in underway replenishment, as illustrated in figure 7-1. High-pressure areas will be attracted toward low-pressure areas, but high-pressure areas repel each other. Much like weather systems, high-pressure areas dominate the direction of water flow, and low-pressure areas are subject to the resulting movements. These forces cause ships to yaw about their pivot points and increase the risk of collision.[3]

Consider this effect as the ship makes its approach on an oiler, as illustrated in figure 7-2. As the bow crosses the stern of the oiler, the high-pressure area of the ship's bow will move toward the low-pressure area on the oiler's stern, but as it gets closer to the hull, its own high-pressure area will push the oiler's stern away.

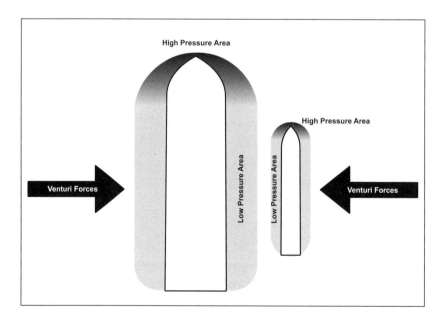

FIGURE 7-1   Venturi Forces while Alongside

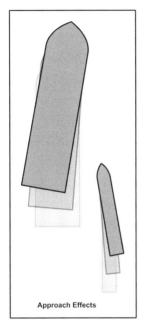
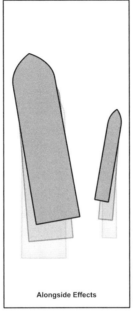
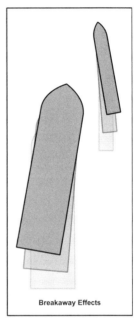

Approach Effects   Alongside Effects   Breakaway Effects

FIGURE 7-2   Venturi Forces throughout the Underway Replenishment

This force will cause the oiler's bow to pivot to the right and toward the ship. This movement of the ship's bow toward the oiler will continue as the bow moves up the oiler's side. As soon as the ship's bow reaches the oiler's bow, the opposing high-pressure areas will push the bows apart, causing the oiler to pivot to port and the ship to pivot to starboard. This opposing pivot will push both sterns toward each other. During the breakaway, the high-pressure area at the oiler's bow will move toward the low-pressure area on the ship's port side, but as the oiler's bow gets closer to the ship's hull, its high-pressure area will push the ship's stern away. This interaction will cause the ship to pivot to port and toward the oiler. Because each of these situations tends to cause the ship and oiler to pivot toward each other, it is clear that Venturi forces are a considerable hazard to ships operating alongside.[4]

As a point of reference, sixty feet of lateral separation is generally accepted as the tipping point between concern and danger of collision. Venturi forces are four times greater at sixty feet than at 120 feet, and these forces increase exponentially with every foot closer. Once the ship is inside sixty feet, it will be very difficult to separate from the oiler. In this situation, the ship cannot simply use maximum speed to escape, because increasing speed will increase the velocity of water along the hull, which will in turn increase the forces pushing the ships together. Venturi forces increase exponentially with speed, as they do with lateral separation; they are four times stronger at twenty knots than at ten knots.[5]

Six Shiphandling Actions

It is helpful to consider how to control the ship through the six shiphandling actions, which were introduced in Chapter Three, on waterjet vector management, and have been employed throughout subsequent chapters. To be sure, shiphandlers have fewer controllable forces at their disposal when the ship is moving ahead at thirteen knots during an alongside replenishment than they do in pierwork. Of the five controllable forces—waterjets, thruster (*Independence* variant only), tugs, anchors, and mooring lines—the conning officer alongside an oiler really only has the waterjets available to control the ship. This means that the shiphandler must use forces on the stern to move the bow, and the list of six shiphandling actions are a useful way to remember the interaction between the waterjets and bow. For example, consider the six shiphandling actions when alongside the oiler's starboard side as given in table 7-1, noting that all ship movements are relative to the oiler.

As discussed in connection with pierwork, the novice shiphandler should prepare a list of the six shiphandling actions beforehand for quick reference during the evolution. This list will be useful if the ship strays into danger and swift action is required. In particular, the conning officer may not instinctively know the action required to save the stern from contacting the oiler, because that action may seem counterintuitive. If Venturi forces push the stern to port toward the oiler, the instinctive action may be to turn away from the oiler—to starboard. Turning to starboard, however, will cause the stern to swing to port—toward the oiler. With the

TABLE 7-1   Six Shiphandling Actions When Alongside, All Movements Relative to the Tanker

SIX SHIPHANDLING QUESTIONS	SIX SHIPHANDLING ACTIONS
1. What makes the stern go to starboard?	Turn to port
2. What makes the stern go to port?	Turn to starboard
3. What makes the bow go to port?	Turn to port
4. What makes the bow go to starboard?	Turn to starboard (but must have enough room between ship and tanker for the stern to swing to port)
5. What makes the ship move ahead?	Increase thrust (one or more throttles)
6. What makes the ship move back?	Decrease thrust (one or more throttles)

six shiphandling actions prepared ahead of time, the conning officer will know that if the stern needs to move to starboard, the waterjets must be turned to port.

Additionally, linking the movement of the bow to the stern makes the conning officer think through the cost of such a maneuver. Looking forward during one moment during a replenishment might cause the conning officer to think that the bow must move right, but the six shiphandling actions would force the officer to look aft before maneuvering the ship. In this case, the table would serve as a prompt to check the lateral separation on the stern before making an adjustment to move the bow.

Autopilot and Manual Steering

Autopilot can be a useful tool when conning alongside for underway replenishment, particularly on a minimally manned ship. Employing autopilot not only eliminates the watch-bill requirement for a master helmsman but provides a replacement that is not subject to fatigue during extended replenishment alongside or high-tempo operations. The use of autopilot, however, does not relieve the requirement for someone to mind the helm. Although this duty does not need to be performed by a dedicated watchstander, someone must be designated to ensure that autopilot remains engaged and is steering the ordered course. Otherwise, the first indication that autopilot has dropped off the line may be a drifting heading.

There may be circumstances where autopilot is unavailable or environmental conditions make the commanding officer feel more comfortable with a human at the helm. In any case, the advanced shiphandler should know how to control the ship manually during underway replenishment. Alongside replenishment can be conducted with any engine configuration that provides sufficient speed, but the ship's maneuverability and responsiveness can hinder the conning officer when making fine adjustments while alongside. Dampening the waterjets' maneuverability can help achieve the finer control needed to stay in position.

For the *Freedom* variant, finer control can be achieved by steering with only the port or starboard steerable waterjets. The main advantage to this technique is that angle changes in one steerable waterjet will produce half of the effect of two waterjets together. Of course, in an emergency situation both combinators should be used to leverage maximum maneuverability until clear of danger. The *Independence* variant provides even finer control of heading when driving manually, since each waterjet has its own combinator. Using just one waterjet to steer effectively reduces heading changes to one-fourth of four waterjets together, allowing the helmsman to swing the waterjet as much as thirty degrees to either side without radically altering the ship's course. Again, in an emergency situation, all four waterjets should be used to produce maximum maneuverability until clear of danger.

Standby Station

In the sequence of an underway replenishment, the term standby station is a misnomer, because the shiphandling team is doing more than simply "standing by" to commence the approach. Maintaining the ship alongside an oiler may seem as simple as ordering the same course and speed as the oiler; however, the dynamic ocean environment and the inherent margin of error in equipment and operators produce fluctuations that require an attuned and responsive shiphandler. Even though the oiler will have notified the ship of the refueling course and speed, once the ship is in standby station—between three and five hundred yards on the starboard side of the oiler—the conning officer must adjust the course and speed until the ship is actually matching the course and speed of the oiler. The conning officer will use the agreed course and speed as a starting point but will make small adjustments to find the combination that most closely matches the oiler's movement.

Before the ship makes its approach, the shiphandler must first determine the intended lateral separation. As a rule of thumb, refueling should occur at a distance between 140 and 180 feet. Distances greater than 250 feet risk unseating the fuel probe, and any advantage gained from closing within a hundred feet would not be worth the risk of tempting Venturi forces. Just because the ship intends to refuel at say 160 feet, however, does not mean that the ship must make its approach at this lateral separation. The safer strategy is to make the approach wider and then move closer when alongside, because that reduces the risk that Venturi forces will affect the ship during the approach. A two-hundred-foot lateral separation is sufficient to prevent Venturi forces from hijacking the approach. Once alongside and the ship has settled longitudinally and the hull pressure areas are more closely aligned, the shiphandler can carefully and deliberately reduce the lateral separation to within 180 feet before connecting to the oiler.

At standby station prior to commencing the approach, there are several ways to determine if the ship is at the proper lateral separation. The minimal manning of this ship class requires the shiphandler to rely on automated systems for simple tasks, and getting the ship into position for underway replenishment is an example. The ship's Automated Radar Plotting Aid (ARPA) allows the conning officer to create parallel index lines that align the ship for its approach. A parallel index line is a computer-generated line on the ARPA display that is parallel to the ship's heading and offset by a certain distance. In this case, the conning officer can create a parallel index line that is parallel to the oiler's refueling course and offset from the ship toward the oiler by the desired lateral separation for the approach. As this line is displayed in front of the ship on the radar screen, the shiphandler is able to determine whether the ship is in position, as illustrated in figure 7-3. If the line bisects the radar return from the oiler, the lateral separation is too close. If the line does not intersect the radar return, the lateral separation is too far. The ship is in position when the line runs tangent to the oiler's radar return. It is important to remember that the radar antenna is centerline on the ship, and the ship's beam must be accounted for in calculating separation. This is particularly important with the *Independence* variant, and its 104-foot beam. Consequently, if the conning officer aims to come alongside at a lateral separation of two hundred feet on the *Independence* variant, the parallel index line should be set for 250 feet to account for approximately half of the ship's width.

The other task while at standby station is to match the oiler's speed. As previously mentioned, the oiler will establish a refueling

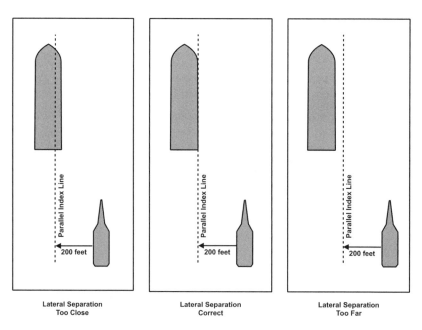

Lateral Separation
Too Close

Lateral Separation
Correct

Lateral Separation
Too Far

Figure 7-3　Parallel Index Line While at Standby Station

course and speed, customarily thirteen knots. The oiler's actual speed may vary slightly due to factors such as engineering-plant margin of error, speed-sensor margin of error, amount of rudder required to maintain course, and increased sea state. The predominance of GPS for speed measurement has lessened the impact of these factors, but in combination they still result in variable speeds.

Of note, thirteen knots is the fleet-wide standard refueling speed, because this speed allows controllable-pitch-propeller ships to maneuver alongside with propellers at 100 percent ahead pitch, controlling forward and aft motion by rpm changes alone. While this ship class does not have reversible-pitch propellers, the principle still applies, as thirteen knots will allow the waterjet reversing plates to remain open at 100 percent ahead. Still, one characteristic of the *Freedom* variant may point to a different speed as a better option—that is, the boost jets engage and disengage around thirteen knots, complicating efforts to maintain this speed alongside. To eliminate this factor, the ship can request a refueling speed at eleven knots, which will allow the ship to conduct the refueling alongside without engaging the boost jets.[6]

In attempting to match the oiler's speed while in standby station, the conning officer will consider both the ordered refueling speed and the speed as indicated on radar. The ship can also contact the oiler to ask for its GPS speed over ground. This combined data should give the conning officer a starting point. When in standby station, the shiphandler will order that speed and then watch the distance to the oiler carefully. If the ship is closing the oiler, the conning officer should reduce speed in half throttle increments; if the oiler is opening, increase speed in half throttle increments. The distance to the oiler can be determined by radar, but an even more accurate method is by laser range finder.

Once the conning officer finds the closest speed, those throttle settings should be recorded as the benchmark for the ship while alongside. For example, if the ship is neither opening nor closing the oiler with the throttles at T4, then T4 will be the first ordered throttle setting alongside. The shiphandler will likely need to make frequent adjustments when alongside to settings both greater and less than T4, but this one will serve as a reference point for closely matching the oiler's speed.

The Approach

When the shiphandler is satisfied that the ship has matched the course and speed of the oiler at standby station, the ship is ready from a shiphandling perspective to commence its approach. While approaching, the conning officer should periodically check the lateral separation to ensure that the ship is not sliding laterally closer to the oiler. The parallel index line is an excellent tool for this purpose as well. Again, if the line begins to bisect the radar return, the lateral separation is decreasing, and the conning officer must alter course to starboard to create more separation. If the line does not intersect the radar contact at all, the lateral separation is increasing, and the shiphandler must alter course to port to decrease the separation. The goal, of course, is to keep the line tangent to the oiler's radar return.

While the parallel index line in ARPA is a very effective automated method, other means are available to check lateral separation and should be used as additional sources of information. The time-tested radian rule will always ensure safe separation during the approach. The radian rule is based on the mathematical certainty that the lateral separation is equal to the distance in yards divided by sixty, times the angular offset between the oiler's bearing and its heading, given that it is steering the refueling course. For example, if the refueling course is 270 degrees true and the oiler's starboard side is observed at 262 degrees true from the ship in standby station five hundred yards astern, the lateral separation is $(500 \div 60) \times 8 = 66.67$ yards, or two hundred feet.[7]

Most novice shiphandlers will be too absorbed in the approach to complete these calculations in stride, so the more practical method is to build a precalculated series of expected angle offsets that serve as checkpoints during the approach. For example, consider the case above, where the refueling course is 270 degrees true and the intended separation is two hundred feet. During the approach, the shiphandler should expect to see eight degrees of angular separation at five hundred yards astern, ten degrees at four hundred yards, 13.3 degrees at three hundred yards, twenty degrees at two hundred yards, and forty degrees at one hundred yards. In preparation for this approach, the shiphandler would prepare a reference card that reads as follows:

RANGE	PREDICTED	OBSERVED
500 yards	262 degrees true	_____
400 yards	260 degrees true	_____
300 yards	256.7 degrees true	_____
200 yards	250 degrees true	_____
100 yards	230 degrees true	_____

This reference card will effectively serve as a glide-slope indicator, such as naval aviators use to check their landing approaches. If the oiler bears 261 degrees true when the ship is four hundred yards astern, the ship is inside the two-hundred-foot approach line, and the conning officer must alter course to starboard to increase the lateral separation. If the oiler bears 259 degrees true at four hundred yards astern, the ship is outside the two-hundred-foot approach line, and the conning officer must alter course to port to decrease the lateral separation. The shiphandler should note on the card which way to turn, because there is little time for analytical thinking during the approach. The card should note:

> If right of predicted, turn right.
> If left of predicted, turn left.

As with the six shiphandling actions, the benefit of this tool is that it can be prepared ahead of time and does not require additional watchstanders. The conning officer only needs a laser range finder and a pelorus to verify lateral separation during the approach. When taking bearings to the oiler, the conning officer should remember to take bearings always from the port-side pelorus to the tangent of the starboard side of the oiler, because lateral separation is measured between the ship's port side and the oiler's starboard side.

Now, consider how the conning officer will manage longitudinal alignment during the approach. Let's say for this example that T4 from all four engines makes about thirteen knots and holds a steady range to the oiler in standby station. To begin the approach the conning officer would order twenty knots, which is fast enough to close the oiler at a steady rate but not so fast as to risk overshooting. For the *Independence* variant, which can control engines individually, the conning officer would order the approach speed (also referred to as stationing speed) by putting the diesels ahead T10 to produce about twenty knots. The benefit of this technique is that leaving the gas turbines in place at T4 increases the likelihood of being able to reestablish quickly the same speed achieved at standby station.

It is also useful for the shiphandler to calculate surge distance prior to the underway replenishment, noting exactly when the ship should reduce speed from twenty knots to thirteen knots. This mark should be referenced from the bridge, right where the conning officer is located. The surge distance for this ship class is about forty-five feet for every knot of speed change, so as illustrated in figure 7-4, when the ship is making its approach at twenty knots (position A), the conning officer must anticipate that the ship will surge 315 feet between ordering thirteen knots and finally settling at that speed. If planning to line up the bridge on the oiler's Station Five in order to align the ship's receiving station with the oiler's Station Seven (position C), the conning officer will reduce speed when the bridge aligns with a point 315 feet astern of Station Five (position B). In this example, this point would be the oiler's stern, so as soon as the bridge passed the stern, the conning officer would order the diesel throttles to T4.

As the ship's approach speed bleeds off and nears the alongside speed established at standby station, the shiphandler will observe where the ship settles relative to the oiler. If the bridge is behind the intended mark, the conning officer can order the engines ahead

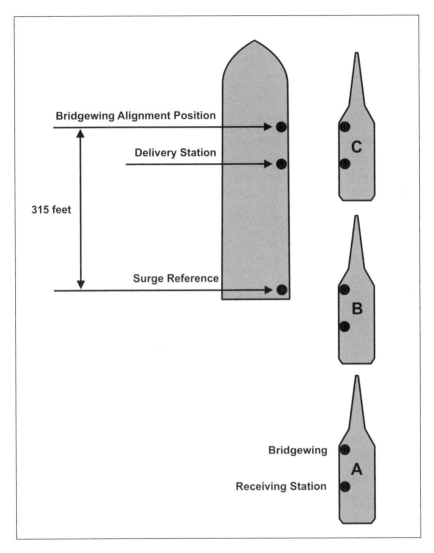

Figure 7-4   When to Reduce Speed on the Approach

T5 to move swiftly into station and then return to T4 when on the mark. If the bridge is ahead of the intended mark, the shiphandler can order T3 to drop back and then return to T4 when in position. These small adjustments will be effective moving slightly forward or aft, but larger adjustments will require more substantial action. The by-half technique described in Chapter Five is particularly useful in these cases. If the bridge settled considerably ahead of its intended mark, the conning officer could drop to T2. As the ship begins to slide back, the conning officer orders T3, then T3.5, and finally T4 once in station.[8]

This approach method above—coasting into position by reducing from twenty to thirteen knots and allowing the ship to surge into place—has become the accepted practice across the fleet, but the introduction of waterjet propulsion has made it possible to return to a more expeditious and precise method that has been nearly abandoned by propeller-driven ships. The braking method allows the ship to maintain the approach speed until quite close to the alongside position by applying astern propulsion to reduce relative speed quickly to zero.[9] Over time, this method has been discarded because of the effect of reversed propulsion on steerageway, but the adoption of waterjets allows the shiphandler to apply braking power while maintaining steerageway.

This method is particularly attractive on the *Independence* variant, where all four engines can be controlled separately, allowing the shiphandler to back down with selected engines while maintaining steering with the remaining ahead engines. To employ the braking method, the conning officer will position the ship in standby station and adjust the throttles to match the oiler's speed. Once again, for this example, T4 on all four engines matches the speed. When ready to make the approach, the conning officer will order both gas turbines ahead T8, which will produce about twenty-five knots. Just prior to the bridge reaching its alongside lineup, the conning officer will order both gas turbines back T8 to apply braking, and the relative speed will drop quickly, as the helm continues to steer with the ahead diesels. When the relative ahead motion is nearly stopped, the shiphandler will order both gas turbines ahead T4. In the balance of art and science in shiphandling, the conning officer will develop a sense of when to apply braking and when to return to the refueling-speed throttle setting. Controlling relative motion with a backing bell will become as intuitive as applying astern propulsion when stopping the ship's forward motion during a docking evolution.

Switching the stationing-speed engines from gas turbines to diesels provides another option for the shiphandler, one involving a slower stationing speed. When making the approach, the conning officer will order both diesels ahead T8 and keep the gas turbines at T4, which produces about twenty knots. Just prior to lining up, the shiphandler will order both diesels back T8 to apply braking while the helm steers with the gas turbines, and then order both diesels ahead T4 when the relative ahead motion is nearly stopped.

The braking method can also be used with all four engines to achieve even greater stationing speed and braking responsiveness. When commencing the approach, ordering all engines ahead T8 will produce about thirty knots stationing speed, and braking with all four engines back T8 will stop the ship's forward relative motion very quickly. Just as in the other configurations, setting the waterjets to T4 at the right moment will promptly put the ship on a steady refueling speed.[10]

The versatility of waterjets provides many other combinations, but the above options offer three distinct stationing speeds: twenty, twenty-five and thirty knots. The uncontrollable forces and tactical situation will dictate which stationing speed is required. The braking approach may be new to some shiphandlers, and concern for losing steerageway could dissuade some from attempting it, but experimentation both at sea and in the simulator has repeatedly shown that the ship does not lose its heading. Even in the case of braking with all four engines, the backing bell is held for only about thirty seconds, and maneuverability is not lost. Still, doubtful commanding officers should practice this approach in the simulator at first, then on the ship in open ocean to watch the heading behavior. If these doubts persist, this maneuver can be practiced alongside an oiler at a safe distance of three hundred feet or more. Duty oilers are generally happy to accommodate commanding officers who have the foresight to practice shiphandling before they go alongside for refueling.

Before moving on to alongside shiphandling, consider one more thought about the approach. There is an old adage that golf is a "game of misses." The philosophy is that a player rarely hits the ball exactly in the intended direction; it will inevitably stray off course in some manner based on the player's ability or prevailing environment. "Good misses" are those that avoid disaster, and "bad misses" are those that find it. For example, if the player faces tall grass on the right and a lake on the left, missing to the right would be considered a "good miss," but hitting into the lake on the left would be disastrous. Therefore, aiming a bit to the right of the center would be the smart play, because a player can still score well with a "good miss."

This adage serves as an ideal analogy for shiphandling when making a replenishment approach. Driving the ship into the oiler would be disastrous, but coming in wide right would be a "good miss." Once alongside, the shiphandler can simply steer slightly left of the refueling course, and the ship will promptly achieve the correct lateral separation. Even if the ship does not actually strike the oiler, missing the planned lateral separation to the left creates an increasingly hazardous situation. Venturi forces increase exponentially with proximity, so every foot closer to the oiler exponentially increases the risk of collision. Missing far enough left will cause the conning officer to battle Venturi forces simply to keep the ships apart. So, if the goal is to refuel at 160 feet, aiming to the right for two hundred feet of lateral separation will set up the conning officer for a "good miss."

Shiphandling Alongside

Shiphandling when alongside is a relative-motion problem, wherein the shiphandler is primarily concerned with two factors: lateral separation and longitudinal position. For lateral separation, if the ship is too far from the oiler, the refueling probe will disconnect; if the ship is too close, Venturi forces will push the sterns toward collision. As for longitudinal position, the delivery and receiving stations must be as closely aligned as possible to allow the hose line to lie perpendicular to the two vessels. The conning officer will spend nearly all of the hour or more alongside managing these two factors.

As in many other stressful shiphandling evolutions, the conning officer can become overwhelmed with information, so some prioritization is in order. Chapter Five discussed the three gauges for pierwork, and their principles apply when alongside the oiler as well. The three gauges for pierwork—heading relative to the pier, longitudinal speed relative to the pier, and lateral closure relative to the pier—work nearly the same in underway replenishment. In this case, the shiphandler should continually check heading relative to the oiler, longitudinal speed relative to the oiler, and lateral closure relative to the oiler.

At the most basic conceptual level, the ship can be considered as either driving toward the oiler or away from it. The oiler will steer a certain course, and maintaining the same heading as the oiler should produce zero lateral motion between the two. While margins of error in equipment and operators will cause the shiphandler to doubt it, this principle serves as a good rule of thumb to visualize the effect of course changes or a fluctuating heading on lateral separation. If the ship is steering left of the oiler's course, it is driving toward the oiler and reducing lateral separation. If the ship is steering right of that course, it is driving away from the oiler and increasing lateral separation. Applying this geometric fact in a more prescriptive way, steering left of the refueling course will decrease lateral separation, and steering right of this course will increase lateral separation. Keeping mental track of this gauge will allow the conning officer to anticipate the next measurement of lateral separation, which is very useful in the dynamic ocean environment. For example, if the helmsman or autopilot is struggling to maintain course and is spending more time left of course than right, the conning officer should expect the lateral separation to shrink and could preempt this change by ordering the ship to come right.

Maintaining the proper separation from the oiler is critically important to preventing collision. If there is one rule in alongside refueling that stands before all others, it is that the shiphandler must always know the distance to the oiler and direction of change. Knowing the second part—direction of change—is imperative, because ninety feet and closing is starkly different from ninety feet and opening. There are several ways to judge this distance, including the phone-and-distance line and seaman's eye, but the most accurate means is by laser range finder. When using this tool, shoot directly perpendicular to the oiler to ensure the truest measurement of lateral separation. Keep in mind that significant course deviations from the refueling course will cause lateral separation at the stern to be different than at the bridge. For example, if the ship is steering five degrees right of the refueling course, the stern will be fifty feet closer to the oiler than the measurement from the bridge indicates.

When the ship is in position, the goal is to maintain zero relative motion longitudinally with respect to the oiler. The best way to gauge this relative motion is to find a "range" on board the oiler, similar to a navigational range marking the center of a channel. This range could be formed by port and starboard refueling rigs, whip antennas, or any two objects that line up when the ship is in position. Watching the movement of these two objects on the oiler will permit the shiphandler to evaluate relative motion quickly. Additionally, the shiphandler should identify the forward and aft points on the oiler that mark the acceptable range of longitudinal motion as the ship slides forward and back relative to the oiler.

The conning officer should provide ordered courses while alongside rather than take the waterjets "in hand" (meaning, steering by waterjet angles). Forces are acting on the ship that are too numerous and interrelated for the conning officer to process: wind, seas, Venturi forces, and tension from the replenishment rig.[11] Taken together, these forces may require considerable waterjet angles to maintain the refueling course, and autopilot is making continuous calculations to determine this angle. If a helmsman is directed to drive manually rather than using autopilot, it may be helpful to set one waterjet left five degrees to provide a constant lateral force that offsets Venturi forces and holds the stern away from the oiler. This arrangement will prevent the helmsman from requiring significant angles on the other waterjets to maintain the ordered course.

The conning officer should take the waterjets in hand only during an emergency and in these instances should be aware that centering the waterjets could result in the stern moving swiftly toward the oiler. If circumstances permit during the emergency, the conning officer should take a few moments to observe the waterjet positions before taking them in hand, to determine the baseline waterjet angle to maintain course. For example, autopilot may require five degrees of left rudder to counteract Venturi forces alongside, so if the conning officer wants the ship to steer straight ahead, the first order with the waterjets in hand should be "Left five degrees rudder."

There are several ways to control the ship's longitudinal position. Some experienced shiphandlers prefer to choose specific shaft rpm to control increases and decreases in speed more finely, but this method can result in the helmsman "throttle hunting" for a specific setting. Excessive throttle hunting, as noted in an earlier chapter, can complicate the conning officer's effort to determine the effects of the last order given. For example, if the shiphandler orders an increase from 280 rpm to 290 rpm, the helmsman might spend the next minute adjusting from 280 rpm to 300 rpm, to 285 rpm, to 295 rpm, and finally find 290 rpm. Throttle hunting confuses the relative-motion evaluation for the conning officer as the speed subtly increases and decreases.

A simpler and more practical approach is to use throttle settings to change the ship's speed, settings similar to those used during pierwork. For example, if the throttle is set to T4 and the ship begins to fall back relative to the oiler, the conning officer can order T4.5, with the understanding that the helmsman will take just a handful of seconds to adjust the throttle and then leave it alone. The conning officer should expect that this method may result in throttle settings slightly above or below the ordered setting, but the loss in accuracy is overcome by the gain in an immediate assessment of its shiphandling effect. In the end, the ship does not need to remain in exactly the same position; it can move forward and back a certain distance without any effect on replenishment operations. Precision has diminishing value once the refueling rigs are nearly aligned, but the shiphandler's clear assessment of relative motion is imperative for safe alongside operations.

The shiphandler can use several additional options to control speed more finely while maintaining the simplicity of this method. For the *Freedom* variant, the conning officer can adjust speed on just the port or starboard combinator by a half throttle setting and leave the other at a constant setting. This configuration will have the effect of changing the speed by one-quarter throttle setting, since only one of two levers is moved by a half setting. On the *Independence* variant, where there is a throttle for each of the four engines, the conning officer can exert even finer control. Altering one of four throttles by a half-setting effectively changes the speed by a one-eighth throttle setting. To be sure, the type of engine used for adjustments will make a difference in how much the speed is affected, since changing speed on the diesel engine by a half-setting will have a smaller effect than the same adjustment on the gas turbine. Using one of the diesel engines for speed adjustments will provide the finest possible control for longitudinal position.

While alongside, there is an interrelationship between lateral separation and longitudinal position that is important to consider. Altering course to either side of the refueling course will reduce thrust in the direction of the refueling course by a certain amount and cause the ship to fall back. If the ship increases speed to maintain its position and subsequently returns to the refueling course, the shiphandler should be aware that the return of all thrust to the direction of the refueling course will cause the ship to move forward, and the ship will need to decrease speed. The advanced shiphandler can anticipate these movements and make preemptive adjustments to speed when changing course, but the effect on longitudinal position is so slight that the novice shiphandler can manage these two gauges separately: alter course to control lateral separation first, and then, second, alter speed to account for changes in longitudinal positioning.

Breakaway

During the breakaway, the most important task is to create distance swiftly between the ship and the oiler. This is accomplished by quickly gaining speed ahead and slowly opening lateral separation to starboard. Ordering all throttles to T10 will build speed swiftly, but using an aggressive waterjet angle or ordering substantial course changes risks swinging the stern into the oiler. As a rule of thumb, adjusting the ship's course by one degree to starboard will swing the stern ten feet to port.[12] Even if the rate of course change is not enough to force the stern into the oiler, moving it too quickly could tempt the Venturi forces. If the lateral separation is 120 feet and the ship's heading is ordered right by six degrees, the stern will swing to port by sixty feet, which puts the hull within the sixty-foot tipping point between concern and danger of collision. The conning officer should avoid changing course by more than five degrees at a time, in order to keep the stern from swinging more than fifty feet toward the oiler. Course changes can be made more conservatively, in two-degree increments, but keep in mind that autopilot cannot be ordered to "come smartly" to a new course—the shiphandler cannot control how quickly autopilot responds.

The breakaway maneuver is another opportunity for the conning officer to weigh the merits of art versus science in shiphandling. The scientific approach would dictate a structured breakaway in which the ship's heading is adjusted incrementally until the bow is well clear of the oiler, but a shiphandler with an attuned seaman's eye can maintain better control over the ship with the waterjets in hand. In this case, the conning officer must remain focused on the separation between the stern and the oiler. Ordering the waterjets right five degrees will promptly move the stern in the correct direction; if the stern appears to be swinging too quickly, ordering the waterjets amidships will ease that movement, particularly if throttles are set at T10 during the breakaway. Only in situations where the stern is still moving toward the oiler should the waterjets be turned left, and then just long enough to check that movement. In the worst case, where the stern is within sixty feet of the oiler, substantial left rudder might be necessary to combat Venturi forces.

Emergency Shiphandling Alongside

Despite the best shiphandling, engineering casualties can put the ship in danger of collision. For example, the oiler can lose steering control, with the rudder stuck either left or right. Lateral separation, ironically, will reduce whether the oiler's rudder is stuck left or right. If the rudder is stuck hard right, the oiler's bow will turn toward the ship and reduce lateral separation forward; if the rudder is stuck hard left, the oiler's stern will swing toward the ship and reduce lateral separation aft. The same casualties can occur on the receiving ship as well, but for this ship class the maneuverability of the waterjets provides many options to maintain steerageway. The shiphandler's primary objective is to keep the ship clear of contact with the oiler long enough to disconnect the rigs safely prior to breaking away, but damage to the refueling rigs is always preferable to contact between the two hulls. If any contact happens, the stern should always be kept clear to preserve the ship's ability to maneuver.

In all cases, the most critical factor is relative motion between the ship and oiler. An ideal maneuver would maintain zero relative motion, the two hulls neither opening nor closing, until the refueling rigs can be safely detached. If this outcome is not possible, slowing the relative motion to open lateral separation slightly might provide enough time to detach the rigs. In no circumstance should the ship be allowed to close within sixty feet of the oiler, at which point Venturi forces can take over and push the ships together.

Shiphandling during alongside emergencies, then, is no time for subtle actions. The shiphandler must take decisive and quick action to keep the ship under control and away from the oiler. First, the conning officer should take the waterjets in hand. Only the shiphandler, with eyes on the distance between the oiler and ship, can fully appreciate the urgency of action. Ordering the helm to "come right" or "come left" does not sufficiently convey the rate of turn required, so the conning officer should give specific rudder orders to exercise the closest possible control over the ship. Second, the by-half technique is most appropriate during these types of emergencies. Any momentum toward the oiler must be arrested immediately, and the most effective way to halt this movement is aggressive action.

Oiler's Rudder Stuck Right

If the oiler's rudder is stuck right and its bow is pivoting toward the ship, the conning officer must match the oiler's stern swing to keep the hulls parallel. Taking the waterjets in hand and ordering them to right thirty degrees, the conning officer should look astern and watch the relative lateral motion between the ships. The sterns will open before the initial rudder order, but once the waterjets are maneuvered thirty degrees to starboard, this motion should stop quickly. When the rapid opening rate begins to diminish, halving the rudder angle to fifteen degrees may be sufficient to make the sterns move together, but if not, the conning officer should continue halving the rudder angle until the sterns are moving at the same rate.

At the same time, the ship will move longitudinally ahead of the oiler, even if the oiler's speed remains the same, because the oiler will make a wider turn than the ship. The conning officer must reduce speed to remain alongside; the by-half technique is appropriate here as well. If the conning officer is controlling speed with just one throttle and that throttle has been keeping the ship alongside at T4, the conning officer should order T0 to stop the ship from surging ahead. Once the forward relative motion ceases, the shiphandler should order T2 to prevent the ship from sliding back and thereafter increase speed by half until the oiler's relative speed is matched.

In an extreme emergency, where turning the waterjets right thirty degrees is insufficient to match the oiler's stern swing, the conning officer can toe-out the waterjets full, with the port engine backing and starboard engine ahead. Using the toe-in/toe-out method will provide ultimate control over relative lateral motion between the sterns, but at this point the shiphandler should clear personnel from the rigs and prepare to escape the oiler quickly, even at risk to the rigs.

Oiler's Rudder Stuck Left

While the novice shiphandler may think that an oiler's rudder stuck toward the ship is the most dangerous scenario, in fact a rudder stuck away from the ship is much more hazardous. The oiler's pivot point at thirteen knots will lie about one-third of its length from the bow, so the stern will come at the ship much faster than the bow would. In this case, the shiphandler should again take the waterjets in hand to match the oiler's stern swing. The conning officer should order the waterjets left thirty degrees and look astern to watch the relative lateral motion between the ships. The sterns will be closing before the initial rudder order, but ordering the waterjets over thirty degrees to port should quickly stop this relative movement. Once the rapid closure rate is reduced, halving the rudder angle to fifteen degrees may be sufficient to keep the sterns moving at the same rate, but if not, the conning officer should halve the rudder angle until the sterns are moving together.

Unlike in the previous scenario, the ship will drop back longitudinally relative to the oiler, since the ship is turning in a wider arc than the oiler. The conning officer must increase speed to remain in position, and the by-half technique will work here as well. If controlling speed with just one throttle, which was keeping the ship alongside at T4, the conning officer should order T8 to stop the ship from falling back. Once the astern relative motion dissipates, the shiphandler should order T6 and thereafter continue decreasing the speed by half until the oiler's speed is matched.[13]

As in the previous scenario, using the toe-in/toe-out method is effective in bailing out of a truly dire situation. If turning the waterjets left thirty degrees is not enough, toe-out the waterjets full, with the port engine ahead and starboard engine backing. Again, the purpose of this maneuver is to keep the stern clear of danger, provide enough time to clear the rigs of personnel, and angle the ship so as to escape the oiler quickly.

If in Doubt, Back Out

In many cases, the safest path away from the oiler is ahead and to the right. This is true for both a normal and emergency breakaway. The only difference in the emergency breakaway is that it happens more quickly. In some circumstances, however, the ship does not have a clear path ahead or the ship cannot clear the oiler ahead without endangering the stern. Again, as Crenshaw's first rule for shiphandling succinctly states, always keep the stern away from danger.[14] In these cases, the best escape route may be astern.

Backing down will remove the stern from the situation first and preserve the ship's propulsion even if the bow makes contact. This maneuver will also develop relative separation most quickly, by adding the ship's astern motion to the oiler's forward motion. This ship class can stop quickly, which will produce thirteen knots of relative separation on short order, and all four waterjets can rapidly develop astern motion. Backing with the throttles at T10 and the waterjets centered will pull the ship straight back.

Shiphandlers on propeller-driven ships might be concerned, if the stern had to be maneuvered away from the oiler, about losing steerageway astern, but this ship class has no such constraints. Simply pointing the waterjets in the intended direction will pull the stern accordingly. In addition, by pulling the stern in one direction, the shiphandler can pivot the ship and swing the bow clear in the opposite direction. For example, consider a situation where the vessels are so close that the stern must be maneuvered around the oiler's stern, as illustrated in figure 7-5. Backing with the waterjets right thirty degrees (position A) will pull the stern away from the oiler. Once the stern is clear astern of the oiler (position B), shifting the waterjets to left thirty degrees will pull the stern behind the oiler and pivot the bow to starboard (position C). It is important to remember that autopilot will not steer a course when going in reverse; backing out requires the shiphandler to take the waterjets in hand.

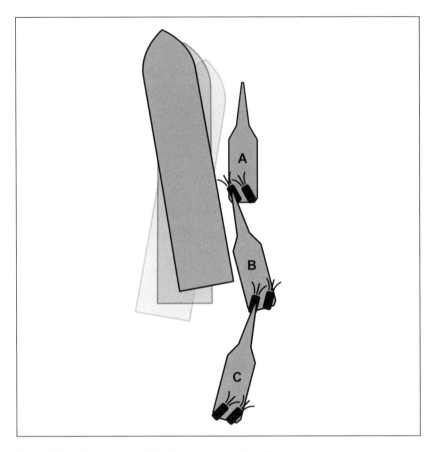

FIGURE 7-5   Maneuvering the Stern around the Oiler

Backing out also provides a quick escape when the receiving ship has lost steering, a more proactive option than waiting for the less maneuverable oiler to take action. The direction in which the waterjets fail is certainly a consideration, but backing with the waterjets centered will pull the ship straight back. If the waterjets are stuck to port, backing will pull the stern to port; the converse is true as well. Still, backing on all four engines will immediately create high relative speed between the oiler and ship that, if there is still sufficient lateral separation when the breakaway commences, the two vessels should separate longitudinally before their hulls are close to contact.

Close-quarters situations alongside can quickly escalate from hazardous to in extremis, so remember this phrase: If in doubt, back out!

SMALL-BOAT OPERATIONS

When alongside the ship, the interaction between the ship, a small boat, and the seas can be hazardous. The novice shiphandler may initially feel at the mercy of the uncontrollable forces in this situation, but in time, will learn to position the seas in the most advantageous location and make best use of the controllable forces to reduce the risk of a mishap. The ship can provide the most protection to the small boat by creating a lee—that is, positioning the ship to protect the boat from the seas. Since the boat deck is on the port side of this ship class, maneuvering to place the seas on the starboard side will create a shadow zone on the port side that produces an area of safe water at the boat deck.

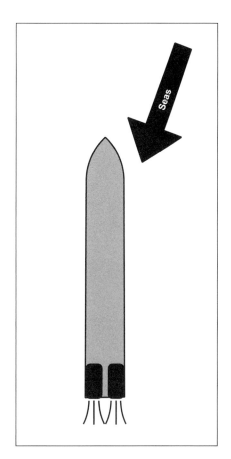

FIGURE 7-6   Simple Lee

Up-Swell Boat Operations

The simplest way to create a lee is to steer a course that places the seas about twenty degrees off the starboard bow, as depicted in figure 7-6. This heading will prevent the seas from directly reaching the port quarter and create a lee roughly from amidships to the stern on the port side. Some waves may refract around the ship, but their height will be greatly reduced compared to the starboard side. This type of lee, a simple lee, will keep the seas on the starboard bow throughout the small-boat operations. The benefit of this type of lee is that the ship can remain on this course indefinitely; the downside, as will become readily apparent below, is that it may not provide the best water alongside the ship in certain sea states.

In higher sea states, placing the seas on the starboard bow may not provide safe enough water alongside, so more active shiphandling may be required. In addition to blocking the seas with the hull, the conning officer can use the waterjets to push the stern against the oncoming seas to create a dynamic lee. This stern movement will effectively smooth out the water, like a trowel drawn across wet cement, as illustrated in figure 7-7. Beginning with the seas ten degrees off the starboard bow, the conning officer should turn the waterjets to port when ready to place the ship's boat in the water or receive a boat alongside. Turning with the waterjets at fifteen degrees will quickly move the stern to starboard, and easing this angle will prevent the ship from swinging across the seas too quickly. The goal is to turn at no more than five degrees per minute to ensure the maximum amount of time before the

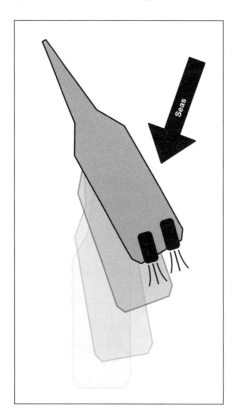

FIGURE 7-7   Dynamic Lee

seas swing past the starboard beam. If beginning with the seas ten degrees off the bow, a five-degree-per-minute turn will allow sixteen minutes of lee time, and a three-degree rate will yield twenty-six minutes. The seas can be allowed to proceed past the beam, but the small-boat coxswain should be notified to expect following seas.

Some alongside operations can take longer than the time permitted by a dynamic lee, and if a simple lee will not provide safe water, the *Independence* variant can use the bow thruster to create a static lee. This maneuver creates the improved conditions of the dynamic lee without constraining the time the boat can stay alongside. By applying the same amount of lateral force on the bow and stern while maintaining five knots of headway, the shiphandler can keep the ship from pivoting and shove the entire length of the ship sideways into the seas. The seas will maintain a constant relative bearing when the static lee is employed, as illustrated in figure 7-8.

To begin this maneuver, the conning officer should first place the seas on the starboard bow at about 030 degrees relative and put the rudder over fifteen degrees to port as described for a dynamic lee. Then, pointing the bow thruster directly into the seas at about 045 degrees relative and setting the throttle at T3, the conning officer should watch the ship's heading carefully. If the ship's heading begins decreasing, the seas will fall off to starboard, and the conning officer must apply more power on the thruster to hold the bow into the seas. If the heading is increasing, the seas will fall off to port, so the conning officer must decrease power on the thruster to prevent the seas from swinging across the bow. The thruster should be sufficient to match the lateral force applied on the stern, but if T10 on the thruster cannot keep up with the stern, the shiphandler should reduce the waterjet angle to slow the stern swing. The goal is to provide equal pressure on the bow and the stern so that the ship slides evenly into the seas, keeping the seas at 045 degrees relative with no relative movement.

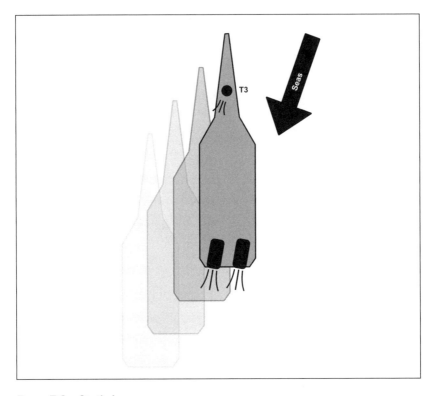

FIGURE 7-8   Static Lee

Down-Swell Boat Operations

There are some circumstances when a down-swell course is more advantageous for small-boat operations. This is particularly true when large swells cause the boat to rise and fall considerably relative to the ship. In these cases, it is safer to drive down-swell to reduce the periodicity of movement. Up-swell seas facilitate a steady course for the ship, but stability for the ship will not make up for the instability of the small boat.

To create a down-swell lee, the conning officer should begin by placing the seas on the starboard quarter, about 160 degrees relative, as depicted in figure 7-9. The most important point is that the seas should not be allowed to outrun the ship. The conning officer must increase or decrease the speed so that the ship either matches the speed of the seas or slightly outruns them. Caution should be exercised when employing a down-swell approach, because quartering seas can cause wallowing that produces erratic stern movement. Unexpected lateral movement can result in either forceful contact with the small boat or rapid separation, both of which are dangerous while people are ascending or descending the ladder. Wallowing can also produce irregular rolls that are especially dangerous for the *Independence* variant, with its trimaran design. If the amah rolls completely out of the water, the small boat could be pulled underneath and suffer catastrophic consequences when the amah descends, so any indication that the bottom of the amah is approaching the

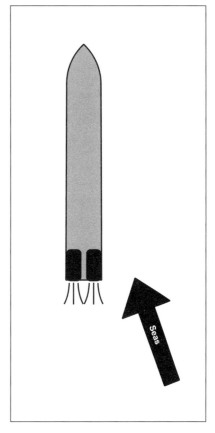

FIGURE 7-9   Down-Swell Boat
Operations

surface should preclude using a down-swell lee.

Wallowing can be reduced by increasing speed or changing the angle of the seas on the stern. Moving the seas to 170 degrees relative will reduce the lateral force applied to the stern, but moving the seas closer to 180 degrees relative may allow refraction or even a direct path for the seas to strike the small boat. In any case, the shiphandler should instruct the boat to remain clear of the ship while different courses and speeds are tried before the boat is brought alongside.

ANCHORING

The adoption of electronic charts has made driving the ship to an anchorage easier than ever. Electronic charts, combined with the accuracy of GPS, enable the shiphandler to see the ship's location and estimate with high accuracy whether the current course and speed will get the ship to the anchorage. In the past, relying solely on visual and radar navigation, it was challenging for the shiphandler to meet Navy requirements to anchor the ship within fifty yards of a plotted anchorage point; today, GPS and electronic charts make anchoring within ten yards fairly routine.

While establishing a head bearing is useful for planning, precise GPS positioning obviates the requirement to approach the anchorage along a specific bearing. This freedom of maneuver allows the shiphandler to weigh other considerations when making the approach—such as traffic density, background lighting, and uncontrollable forces—that may not have been apparent when planning the anchorage. The same holds true for the drop bearing, since the GPS location will tell the conning officer when to drop the anchor. Of course, this freedom of maneuver does not relieve the shiphandler of the responsibility to plan properly a route to the anchorage and to be familiar with visual cues in the area.

For novice shiphandlers, envisioning how to drive the ship to a precision anchorage can be difficult. They can find this evolution overwhelming, particularly when they consider how to orchestrate everything that must occur in coordination with the anchor team. From a shiphandling aspect, however, the maneuver is fairly straightforward. The anchorage point is stationary, so driving to the anchorage is a true-motion problem; the conning officer must simply "collide" with a static anchorage point. This mental image of colliding with the anchorage is particularly useful when making the approach with electronic charts, because the anchorage is reported as a continuously updated bearing and range to the anchorage point. When maneuvering relative to another ship at sea, the shiphandler knows that collision is likely if the contact remains on a constant bearing with a decreasing range. Similarly, the shiphandler will know that the ship is on track for the anchorage when the anchorage point remains on a constant bearing with a decreasing range.

To begin, consider the sequence of events of anchoring without any uncontrollable forces acting on the ship, a situation in which the conning officer would make a long steady approach on a course directly at the anchorage point. The gas turbines are not necessary for anchoring, since two diesels provide sufficient power to maneuver the ship; also, running the gas turbines, even with the throttles set at T0, can produce thrust, which is undesirable when trying to control the ship's speed over the anchorage. When the ship is a thousand yards from the anchorage, the conning officer should slow to ten knots. The speed should be reduced again to five knots five hundred yards from the anchorage; at one hundred yards from the anchorage, the conning officer should order both engines back T2 to stop the ship and build slight sternway. The ship will surge seventy-five yards ahead as it slows to a dead stop. The distance from the hawse pipe to the bridge is about one-third the ship's length, so the anchor will have passed just beyond the anchorage point when the ship has come to a stop.

Slight sternway will draw the hawse pipe back across the anchorage, and the conning officer should order "Let go the anchor" when sternway develops, then allow the chain to pay out ahead as the ship backs away.[15] While this ship class does not have a sonar dome that could be damaged if the anchor chain drifts beneath the keel, it is only prudent to ensure that the chain remains clear of the ship. As the anchor falls and the chain pays out, the conning officer should apply just enough backing power to maintain slight sternway without pulling the anchor free. This may require the shiphandler to back on one engine or alternate between all stop and back T1. As it does in any shiphandling evolution, GPS has its limits in providing accurate speed indications when nearly stopped; there is no substitute for throwing markers over the side, such as wood chips or bread slices, and observing their movement.

Now, consider this same sequence with the uncontrollable forces at work. For example, if the ship is approaching an anchorage on a course of 000 degrees true and the true bearing of the anchorage begins to fall off to port—359, 358, 357—the conning officer knows that the ship is being pushed to the right. The novice might order the ship to steer whatever bearing is reported, in an effort to point the ship at the anchorage, but the uncontrollable forces that caused the ship to drift right will continue to act on the ship, despite the course change. This will cause the conning officer to chase the anchorage point as it falls off and around the ship. In order to put the ship on a constant bearing toward the anchorage again, the conning officer must steer left of the bearing to the anchorage. In this example, the ship must steer at least 356, and once on that course, the conning officer should evaluate the effects. If the anchorage point continues to fall off to port, the ship must steer more than one degree left of the bearing to the anchorage. The conning officer will have found the right course when the bearing to the anchorage is constant and the range is decreasing. To recall the image of "colliding" with the anchorage, the ship will miss the anchorage if it has left or right bearing drift; the ship can arrive at the anchorage only if the anchorage point has a constant bearing and decreasing range. Once this motion is achieved, the difference between the ship's course and the bearing to the anchorage is essentially irrelevant.

The prudent shiphandler, understanding how uncontrollable forces affect the ship, will take these forces into account when planning the approach. Undesired effects from wind, seas, and current can be minimized by placing on the bow whichever uncontrollable force has the largest effect on the ship. For example, if the ship is facing heavy ocean seas from 270 degrees true and light winds from 315 degrees true, the best course would be 270 degrees true toward the anchorage. The conning officer must be aware, in this case, that the winds on the starboard bow will have some effect, but the most impactful uncontrollable force, the seas, will be largely minimized as a shiphandling factor when heading directly into them.

If the geographic constraints of the anchorage area do not permit flexibility in choosing an approach course, careful consideration must be given to how the uncontrollable forces will affect the ship. Approaching the anchorage downwind or down-current, with these forces coming from astern, will allow uncontrollable forces to push on the stern as the anchor is released. Once the anchor takes hold, the pivot point will move forward to the hawse pipe, and the uncontrollable forces will push the ship to the opposite side of the anchorage. This movement will point the ship into the wind or current. Rather than fighting these uncontrollable forces, the shiphandler should anticipate these movements and position the ship relative to the anchorage so that the ship falls into place. For example, as depicted in figure 7-10, if the ship is required to makes its approach on a course of 090 degrees true and the wind and current are coming from 270 degrees true, the conning officer could drive toward a point that is fifty yards to the right (or south) of the anchorage point. Once the bow reaches that point, the conning officer would toe-out the engines and twist the ship to port with the port engine ahead T2 and starboard engine back T2. This maneuver will position the bow over the intended anchorage point and ensure that the stern falls off in the intended direction. When the bow is over the anchorage point, the conning officer should

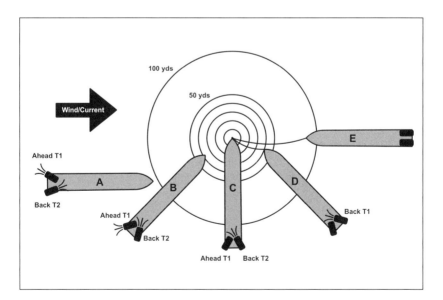

Figure 7-10 Anchoring with Uncontrollable Forces on the Stern

center the waterjets, stop the port ahead engine, ease the starboard backing engine to T1, and let go the anchor when the ship has sternway. The uncontrollable forces should push the ship comfortably away from the anchor. When determining whether to aim left or right of the anchorage point, the shiphandler should choose the side with the maximum room for the stern to swing.[16]

Anchoring with crossing uncontrollable forces is equally challenging, but a shiphandler who anticipates ship movements can stay ahead of the elements. For example, as depicted in figure 7-11, if the ship is making its approach on a course of 000 degrees true and the wind is coming from 270 degrees true, the wind will push the ship to starboard. The conning officer should begin the approach by steering five degrees left of the anchorage. Just as in an approach into the wind, the goal is to achieve constant bearing and decreasing range to an anchorage point. If the bearing drift of the anchorage point is left, the conning officer should alter course to ten degrees left of the anchorage bearing. The key to overcoming crossing uncontrollable forces is to take substantial action early, before the elements put the ship in a position from which it cannot recover. It is easier to back off from aggressive initial action and allow the wind to blow the ship into position than to take radical action later in direct opposition to the wind. Once the bow is over the anchorage, the shiphandler should anticipate that the wind will pivot the ship about the anchor. After letting go the anchor, the conning officer should work the stern toward its inevitable direction. In this example, with the wind on the port side, the shiphandler would toe-out the starboard diesel to expedite the stern's movement relative to the wind and, when the ship is pointing into the wind, center the waterjet.

One final note about anchoring is worthwhile here. When the ship is within one hundred yards of the anchorage and the engines have been ordered astern to stop the ship and build sternway, even the most radical maneuvers will do little to improve the ship's position. Two options remain: drop the anchor or go around again. Many shiphandlers have found themselves twisting the ship in vain, certain that they can recover from a bad approach, only

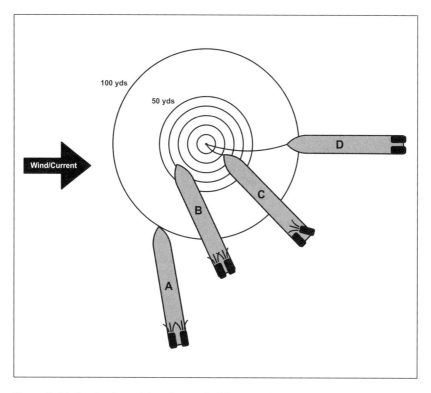

FIGURE **7-11** Anchoring with a Crosswind Approach

to find themselves unmercifully pushed around by the prevailing uncontrollable forces—literally twisting in the wind. If the ship is not close enough to the anchorage, the best choice is to swallow one's pride and go around again. In the end, this option will likely take less time than endlessly twisting in the wind.

In particularly adverse weather, shiphandling does not cease once the anchor is set. Uncontrollable forces producing excessive strain on the anchor could cause the ship to break free of the anchorage. Propeller-driven ships have long used engines to ease the strain on the anchor when the wind or current is forcing them from an anchorage, but this ship class's waterjets provide an added capability. In addition to easing the strain by using forward thrust, the maneuverability of the waterjets allows the shiphandler to control yaw. The wind or current may try to force the ship to yaw both to the left and right of the anchor chain, but the engines can be positioned to counteract those forces.

The goal of using the waterjets to control strain is to allow the ship to ride lightly on the chain straight back from the anchor. The first element—riding lightly on the chain—is accomplished by applying slightly less power than the amount required to overcome the uncontrollable forces, to prevent the ship from surging ahead and thereby prevent the chain from drifting under the hull. Table 7-2 is derived from Crenshaw's guidance on this topic.[17]

In reading this table, the conning officer should use the sustained wind speed to determine the initial amount of thrust for the waterjets. The emphasis is placed on "initial," because the dynamic nature of storm forces requires the shiphandler to adjust the propulsion as winds increase and decrease. If the ship begins moving forward, the conning officer should back off the throttles by a half-setting. If the anchor watch reports heavy strain on the anchor chain, the throttles should be increased by a half-setting. The dynamic nature of these forces will require repeated adjustments. Finer adjustments can be made by adjusting just one engine.

The second element—riding straight back from the anchor—is accomplished by steering the waterjets to maneuver the stern to counter yawing effects. The anchor watch will report how the anchor tends—the ideal condition is an anchor tending at twelve o'clock—but the more immediately useful feedback for the conning officer is the direction of the wind. As depicted in figure 7-12, if the wind is directly on the bow, the ship is riding straight back.

TABLE 7-2   Waterjet Speed Required to Ease Strain on the Anchor, Both Diesels On Line

WIND SPEED	SHIP SPEED	THROTTLE SETTING
35 knots	2 knots	T0.5
50 knots	3 knots	T1
65 knots	4 knots	T1.5
80 knots	5 knots	T2

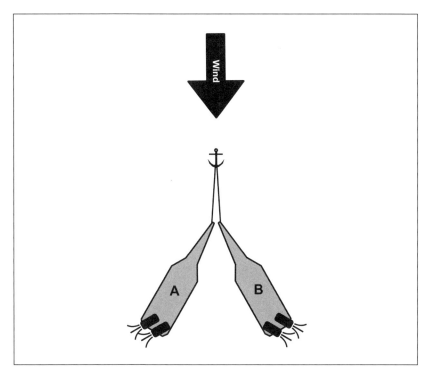

Figure 7-12 Easing Strain on the Anchor Using Waterjets

If the wind is on the port bow (position A), the wind has caused the ship to yaw to starboard and the conning officer should turn the waterjets left to push the stern back to starboard. Conversely, if the wind is on the starboard bow (position B), the wind has caused the ship to yaw to port and the conning officer should turn the water-jets right to push the stern back to port. If the wind is making it too difficult to control the stern laterally, the shiphandler can always use the toe-in/toe-out method to produce the maximum lateral force on the stern. The ship will still require an ahead resultant force vector to ride lightly on the chain, so the shiphandler must apply more power on the ahead engine than on the astern engine. Of course, no matter the waterjet arrangement, the dynamic nature of storm forces will require the shiphandler to check the wind speed constantly and make repeated adjustments.

MAN OVERBOARD

Since this ship class is faster and more maneuverable than propeller-driven ships, the shiphandler can bring the ship back around to a man overboard (a traditional maritime phrase referring to either gender) more quickly, but the maneuver still follows the Navy-wide principles of a recovery. The goal is to establish visual contact with the man as soon as possible and create a lee to launch the boat safely, in such a way that the boat has the shortest distance to travel to pick up the man.[18] The novice shiphandler will drive the ship directly back to the man without much regard for the mechanics of launching the boat, but the advanced shiphandler knows that speeding directly back will not mean much to the man in the water if the ship spends the next ten minutes trying to create a lee to launch a boat.

This is particularly important for this ship class, for several reasons. First, the ship's low draft and large sail area make it susceptible to the wind's effects, so placing the ship on the leeward side of the man could allow the wind to blow it away from the sailor. Second, the ship's lightweight design results in greater pitching and rolling at slower speeds in the open ocean. Excessive movement caused by haphazard placement of seas relative to the ship could cause an even worse problem, since trying to launch the boat without a lee could result in a second man overboard. Shiphandlers spend a lot of time debating which maneuver will most quickly return to the man overboard, but if the maneuver does not produce a lee to safely launch the boat, the maneuver is at best incomplete, and at worst dangerous.

Much like driving toward an anchorage point, man-overboard maneuvers are, by their nature, true-motion problems.[19] The goal is to make the final approach so that the man is on the port side—the same side as the boat deck—and the wind is to starboard. When the man is in sight, the conning officer can quickly map out the approach. This forward thinking will allow the conning officer to use the ship's speed to place the ship on an optimal approach ahead of time, rather than racing to the man and then twisting in the wind. For example, if the ship is on a course of 000 degrees true when the man falls overboard, with the wind from 270 degrees true, the conning officer needs to drive a reciprocal course of 180 degrees true in order to place the man on the port side and the wind to starboard. A Williamson turn would provide the ideal approach. In the same situation, if the wind were from 090 degrees true, the shiphandler would want to approach the sailor on the same course the ship followed when he or she fell overboard, since the wind was already on the starboard side. In this case, a racetrack turn would be more appropriate, since it would return the ship to its original course. On those rare days at sea when the winds and seas are negligible and creating a lee is not a great concern, the Anderson turn and power twist remain the most expeditious methods of returning to the man.

MISSION-BAY OPERATIONS

This ship class is designed to launch and recover waterborne mission-package vehicles astern. The *Independence* variant utilizes an extendable crane that lowers vehicles to the water from its mission bay high above the waterline; the *Freedom* variant has a stern ramp right at the water's edge. Regardless of the method for launching and recovery, the shiphandling fundamentals remain the same.

Launching and Recovering Vehicles

When launching or recovering waterborne vehicles, the conning officer's goal is to provide a steady course and speed with minimum pitch and roll. The ship's course will be most stable when heading into the seas, as the helm or autopilot will have the most control over the stern. In addition, heading into the seas will provide a partial lee astern of the ship, shielding the vehicle from oncoming waves.

With the vehicles directly behind the waterjets, careful consideration should be given to the behavior of the waterjet wash. For the *Independence* variant, the inboard waterjets should be secured to prevent them from disrupting the stability of underwater vehicles or swamping surface craft. This issue is not a concern for the *Free-dom* variant, since the inboard waterjets will not discharge water at the slower speeds used for vehicle launch and recovery.

Although the outboard jets are relatively far from the ship's centerline, the amount of water they discharge can still have adverse effects on the vehicles, particularly if the waterjets are swung left and right to maintain a course into the seas. To minimize the effect of wash across the vehicles, the waterjets can be toed-out—projecting the wash as far from the vehicle as possible—and the conning officer can use thrust to steer the ship. With the waterjets toed-out thirty degrees, increasing thrust on the port waterjet will push the stern to starboard and alter the ship's heading to port. Increasing thrust on the starboard waterjet will push the stern to port and alter the ship's heading to starboard. So, the conning officer should increase thrust on the port waterjet to alter course to port and increase thrust on the starboard waterjet to alter course to starboard.

Keep in mind that the goal during launch and recovery is to maintain both a steady course and speed; increasing thrust on one side will produce an undesired increase in speed. To prevent this outcome, the conning officer can use the *additive technique*, which aims to keep the sum of the two throttle settings constant. For example, if T2 on both waterjets provides a desired five knots of headway, adding the T-settings would total four. No matter what waterjet thrust is required to alter course during this particular evolution, the conning officer should adjust the other waterjet so that the sum equals four. If the heading is falling off to starboard, adjusting the port waterjet to T2.5 and the starboard waterjet to T1.5 will alter the heading to port with little change in the ship's speed. Once the ship is on the correct heading, adjusting both waterjets to T2 will return to the baseline. This configuration can be maintained for hours while launching and recovering vehicles.

Another method of steering, available to the *Independence* variant, involves the bow thruster. Both waterjets can be toed-out thirty degrees and the thruster used to keep the ship on heading. If the ship's heading is falling off to port, the conning officer will direct the thruster to the starboard beam; when the ship is back on course, the shiphandler will stop the thruster and return it to centerline. More simply, the shiphandler could maintain constant propulsion on the thruster and just keep it pointed into the seas, but adding a longitudinal force vector from the thruster may affect the ship's speed. As long as the throttle setting is kept low, however, the thrust will not be powerful enough to alter speed significantly. Employing the thruster at T3 or less will not produce radical changes in ship's speed when swinging the thruster back and forth, and any changes can be quickly corrected by increasing or decreasing the thruster's throttle setting.

While steering into the seas is preferable for a steady course, there are circumstances when a down-swell approach is more advantageous for launching and recovering vehicles astern. This is particularly true when large swells cause excessive vertical relative movement between the vehicle and ship. In these cases, a down-swell course may reduce the swell's apparent periodicity. The ship's speed should match or exceed the speed of the seas, to prevent the swells from pushing the vehicle into the stern. Since a ship with following seas cannot provide an astern lee, the boat coxswain or submersible-vehicle operator may have more difficulty steering. This downside should be weighed as a trade-off against the excessive vertical movement of an up-swell course.

It is important to understand the wake effects occurring behind the ship when vehicles maneuver close to the stern. Barber describes the concept of quickwater as water disturbed by the backing screws on a propeller-driven ship, appearing on the surface as a swirling pattern similar to small eddies.[20] He presents quickwater as a consequence of applying astern propulsion while still moving ahead to stop the ship, but waterjets on this ship class always produce quickwater while using ahead propulsion at low speeds. When the reversing plate is partially open, meaning that less than 100 percent of the discharge is directed astern, some portion of water discharges forward and creates quickwater; this quickwater moves with the ship as the ship proceeds ahead at these speeds. This effect is particularly apparent while the inboard waterjets are secured and the outboard waterjets are toed-out thirty degrees. As viewed from the stern, the outboard waterjets direct astern-discharged water away from the centerline, and a mass of quickwater from the forward-discharged water appears immediately behind the ship and moves with it. This quickwater sits right where the vehicles are launched and recovered.

This phenomenon can be hazardous for vehicles behind the ship, for several reasons. First, the lack of flow across the vehicles' controlling surfaces, such as rudders and fins, will make it more difficult to control them. Second, this quickwater rides in a low-pressure area created by the ship's movement and presents a danger of the vehicle being pushed into the stern by Venturi forces. The force pushing the vehicle toward the ship increases exponentially with decreasing distance; the closer the vehicle, the more pronounced the effect. It is possible to diminish this effect by centering the outboard waterjets to push water into this area, but increased flow presents its own hazards. As described above, the turbulent waterjet wash does not provide the constant flow required for controlling surfaces on underwater vehicles, and it can disrupt the recovery devices used to pull surface craft out of the water.

Cargo Handling

Moving mission-package equipment and vehicles around the ship can be hazardous at sea, particularly in higher sea states. The ship-handler must make every effort to keep the ship as steady as possible to reduce the risk of mishaps. Steering courses into and with the seas will keep the ship steadiest; speed adjustments allow the ship-handler to keep excessive movements under control.

It may be necessary to make course changes while moving mission-package equipment, and the conning officer may do so as long as pitch and roll remain reasonable. To prevent the ship from heeling over too far during the turn, the time-tested "rule of thirty" can be very helpful—the sum of the waterjet angle and ship's speed should not exceed thirty. For example, if the ship is making twenty knots, the conning officer should not allow the waterjet angle to exceed ten degrees. If the ship's speed is five knots, however, the shiphandler can use up to twenty-five degrees of rudder. The exception to this rule arises when the ship is turning through the trough. As the seas present on the beam, the ship may get stuck in the trough, and the conning officer must power through it with maximum rudder and increased thrust. The shiphandler may resume the rule of thirty once the seas are clear of the beam.

CONCLUSION

It is for good reason that underway replenishment commands a large portion of this chapter. The ability to sustain ships at sea is a defining characteristic of the U.S. Navy, and its ability to refuel at sea allows this ship class to operate for many weeks without returning to port. This is particularly true when the ship is engaged in sustained high-speed operations that burn the most fuel. The ship's combat readiness is dependent on its ability to refuel at sea, so the shiphandling team must be ready to take fuel on demand and safely.

As mentioned at the beginning of this chapter, the above shiphandling evolutions have a quality that makes them "special"— that they are infrequent and carry an elevated risk of damage to equipment or injury to sailors. This characterization naturally commands the attention of the shiphandling team during the first several events, but the danger of complacency always looms in the background. As the shiphandling team accomplishes more of these special evolutions, it becomes more proficient and the evolution becomes more routine. On one hand, the shiphandlers will gain confidence that eases the stress of close-quarters maneuvering; on the other hand, however, successfully completing an inherently dangerous evolution can lead the team to think that elevated risk no longer exists. It is incumbent on the more experienced mariners to remind novice shiphandlers that mishaps can occur on any given day and that the team must be prepared to take swift action to prevent a mishap from becoming a disaster.

CHAPTER EIGHT

Open-Ocean Shiphandling

This ship class was designed to operate primarily close to shore, in shallow waters largely protected from the effects of heavy weather in the open ocean. Deployment of these U.S. Navy ships, however, requires transits across vast expanses between the United States and overseas operating stations; also, once in theater the ship will often be required to operate in the open ocean. Whether joining up with an aircraft carrier on the high seas or heading to an at-sea nighttime operating box where there is little traffic, this ship class cannot escape the requirement to operate outside coastal waters.

The ship's lightweight design and shallow draft are the most significant shiphandling factors to consider for open-ocean operations. While this ship class is open-ocean capable, it is not open-ocean optimal. These ships ride high and are easily tossed by the prevailing seas. If left to the mercy of the uncontrollable forces, they will move considerably, so their conning officers must actively drive them in a way that minimizes undesired effects of this movement.

The novice shiphandler might not fully appreciate the dangers of a ship at the mercy of the sea, because warships appear invincible. No warships are unbreakable; even the heaviest and most strongly built ships suffer damage from seas. This damage can occur in the hull framing or superstructure plating. It can occur in combat systems equipment that is open to the elements. It can even occur in internal fittings, such as swinging doors or falling hatches. Ships are designed to sail up to their sea-state limitations, but the dynamic nature of the sea presents hazards that are unpredictable. Sailing ships in hazardous conditions is not unusual, but it should always be accompanied by deliberate risk management. Steaming blindly on a hazardous course without concern for the ship's safety is not the mark of an advanced shiphandler.

Additionally, the hazards that threaten equipment are even more likely to endanger the crew. When rough seas lay into the ship, they can be uneven, and unexpected waves can ambush sailors carrying out the ship's work. A door that is not secured properly can swing open into a sailor standing nearby or swing shut onto a hand holding its "knife-edge" door jamb for support. The possible injuries are too numerous to list, but the lesson is the same as above. Steaming blindly without concern for the safety of sailors is a mark of inexperience.

Finally, the impacts of rough seas on a small crew is far more significant than on a ship with deep watch-standing reserves available. This ship class requires that every sailor stand watch effectively, and rocking the ship unnecessarily during meals or sleeping hours causes lost rest for off-watch crew members. Over time the crew will suffer from fatigue, and these fatigued sailors must continue standing watches, subsequently endangering the ship. The conning officer has a responsibility therefore to drive the ship in a manner that allows off-watch crew members to rest. If the point was not clear in the previous two paragraphs, it bears repeating here. Steaming blindly without concern for fatigue management indicates a failure of the shiphandler to think beyond the bridge.

This chapter is designed to help the conning officer manage this problem, beginning with identifying the direction of the seas and understanding how certain relative seas affect ship movement. Then, it will discuss methods for seakeeping when operational commitments seem to conflict with courses that would provide for the best ride. Finally, this chapter will consider how the shiphandler can use modern forecasting and charts to identify safe-water areas.

KNOWING THE SEAS

Because of the unique challenges of operating this ship class in the open ocean, the conning officer must constantly evaluate the seas and their effect on the ship. Ship movement depends on the relative direction of seas impacting the ship and the apparent periodicity of impact from successive waves. The following discussion will first discuss how the relative direction of the seas will affect ship movements and second how the relative periodicity affects the severity of these movements.

Relative Direction

From a seakeeping perspective, ships move principally in four directions—pitch, roll, yaw, and heave.[1] Pitch is movement along the longitudinal axis, a seesaw type motion, marked primarily by the bow and stern pivoting about the center of the ship. Roll is the lateral movement of the ship, a pendulum-type movement, that pivots on the ship's center of gravity. Yaw is a twisting movement, similar to twisting described in Chapter Three, on waterjet vector management, and in Chapter Five on pierwork, where the ship twists about the pivot point. Heave is level movement of the ship up and down relative to the ocean floor.

Consider the range of relative direction of seas, from bow to stern, and how they will affect these movements. Pitch and roll are inversely related: seas on the bow and stern create maximum pitch and minimum roll, but seas on the beam create minimum pitch and maximum roll. Yaw is induced by seas striking forward or aft of the pivot point; it is maximized by seas at the intercardinal points—045, 135, 225, and 315 degrees relative—and minimized by seas on the bow, stern, and beams. The actual amount of heave is unchanged by the relative direction of the seas, because the ship will rise and fall regardless of their direction, although heave periodicity will change. The closer the seas are to the bow, the more frequently the ship will heave. Seas closer to the stern will induce heave least frequently. Let's examine both the cardinal and intercardinal points to understand better what movement to expect.

Bow Seas

Seas on the bow cause the ship to pitch but yield little roll. Minimizing roll is a benefit, since roll produces the most motion on any type of ship, but the cost of pitching is the risk of excessive bow slamming that could cause structural damage.

Seas Broad on the Bow

With seas at the forward intercardinal points—045 and 315 degrees relative—pitching will be reduced, along with the potential for bow slamming, but rolling will increase. Seas striking off center from the bow and beam will introduce forward of the pivot point a lateral force that causes the bow to yaw away from the predominant seas.

Beam Seas

Of all the possible directions to take the seas, beam seas are by far the most dangerous for the crew. The ship will experience the least pitch and the largest roll, since all of the force is applied laterally. Ships roll more easily than they pitch; the ship will always experience its most violent movement in roll. Even in a transitional state, such as a course change that requires the ship to present the beam momentarily toward the seas, exposure to beam seas can suddenly become prolonged if the ship becomes stuck in the trough.

Quartering Seas

By shifting the seas from the beam to the aft intercardinal points—135 and 225 degrees relative—rolling will be reduced and pitching increased, and the lateral force offset from the pivot point will cause the stern to yaw away from the predominant seas. The critical difference from the effect of seas on the forward intercardinal points is that the lateral force is now being applied on the same side of the pivot point as the waterjets, complicating steering. Additionally, since the pivot point shifts forward when the ship is making headway and the pivot point shifts away from the lateral force applied on the quarter, the yaw induced by stern movement is much greater than from bow movement. These factors cause a slow, uncontrolled stern movement referred to as "wallowing," with the after end of the ship moving sluggishly about the pivot point and returning just as slowly. While not optimal for maintaining course, this situation would nevertheless be preferable to more hazardous ship movements, such as bow slamming from bow seas or excessive rolling from beam seas.

Stern Seas

Like seas on the bow, a following sea will increase pitch and minimize roll. The added benefits of stern seas, however, are that their apparent periodicity is sharply reduced (as explained below), and bow slamming is significantly reduced, if not eliminated altogether. The disadvantage of stern seas is that the waterjets are directly exposed to them. The waterjets could breach the surface, but even if they remain submerged, it will be difficult to maintain course.

Periodicity

Every type of ship, as a characteristic of its design, has a "perfect" rolling period in which the vessel will make a complete roll, regardless of its severity. If the periodicity of the oncoming waves happens to match the ship's rolling period, each wave that impacts the ship will add to the cumulative effect of the preceding waves, and the ship will roll more and more severely until it achieves its maximum design roll. At this point, given that the sea state has not exceeded the ship's design limits, the righting arm will be strong enough to force the ship upright and prevent it from capsizing. The ship will continue rolling to this limit unless the shiphandler takes action to disrupt the synchronization between the ship's rolling period and the periodicity of the seas.[2]

Disrupting this synchronization can occur through either course or speed changes, since the conning officer needs to change only the apparent periodicity, not the real periodicity. Again considered from bow to stern, placing the seas on the bow will shorten periodicity, and placing them on the stern will lengthen it. Of course, increasing or decreasing the ship's speed can achieve the same effect.[3] Shiphandlers can never change the real periodicity of the seas, but by changing course and speed, they can always change the apparent periodicity.

SEAKEEPING

With this appreciation of relative direction and apparent periodicity, the shiphandler can take a more proactive approach toward handling the ship at sea. *Seakeeping* is the ship's ability to handle well at sea, and this ability is based on several factors. Design factors such as length, beam, displacement, stability, and freeboard play a considerable role, but these factors are decided long before the conning officer steps on board. The only factor that is under the shiphandler's control is, fittingly, shiphandling. Ships are not trains fixed to a track but rather maneuverable vessels that operate in a dynamic environment. Shiphandlers have both course and speed at their disposal to change the apparent periodicity of the seas, and they are empowered to dampen the ship's movement in order to care for both the ship and crew.

The above discussion of how the relative direction of the seas affects ship movement and how the relative periodicity affects the severity of this movement should drive the conning officer's approach toward seakeeping. Of the four major ship movements—pitch, roll, yaw, and heave—pitch and roll are the most significant factors for seakeeping. The shiphandler can control heave periodicity but not its severity, and yaw is less about safe shiphandling than about how much rudder is required to maintain course. Focusing on pitch and roll allows the conning officer to make course changes for seakeeping, with the understanding that these two factors are inversely related when changing course but maintaining speed.

The conning officer should evaluate the ship's movement to determine whether pitch or roll must be reduced and assess whether the opposing factor can be permitted to increase. For example, if the seas are at 050 degrees relative and the ship is experiencing excessive roll, the conning officer can alter course to starboard by twenty degrees to place the seas at 030 degrees relative. This maneuver will decrease roll but increase pitch, so it is acceptable only if the ship is not at its limit for pitching. Conversely, if the seas are at 020 degrees relative and the ship is experiencing excessive pitch, with bow slamming, the conning officer can alter course to port by twenty degrees, which would place the seas at 040 degrees relative. This turn would decrease pitching but increase rolling. Again, this maneuver is acceptable only if the ship is not at its limit for rolling. From all this it becomes clear that course changes act as a rheostat for controlling pitch and roll, whereby pitch is reduced by moving the seas away from the bow and roll is reduced by moving the seas toward the bow.

Seakeeping with a following sea is similar, with pitch reduced by moving the seas away from the stern and roll reduced by moving them toward the stern. The principle difference from head-on seas is that following seas can cause wallowing, particularly when they are allowed to overtake the ship. Wallowing can be reduced with a course change alone by shifting the seas closer to the stern, but this type of ship movement is better resolved with speed changes. When running with the seas, the ship will ride the smoothest when its speed slightly exceeds the speed of the seas.

Overall, unless the seas are so great that they risk bow slamming forward or waterjet breaching astern, placing the seas on the bow or stern will provide the smoothest ride for this ship class, simply because those relative angles will minimize rolling. The shiphandler should be careful, of course, not to drive too fast into head-on seas that present a hazard to the bow structure. As mentioned previously, beam seas should be avoided whenever possible. When beam seas are unavoidable, speed changes are the shiphandler's best tool to control undesired ship motion. The goal in altering speed is to change the apparent periodicity, which can be achieved by either increasing or decreasing speed; increasing speed is generally the better option, if available, as this ship class normally rides better at higher speeds. Additionally, if the ship is required to take seas on the beam because a certain leg of a transit cannot be altered, using the maximum speed available will allow the ship to transit the leg as quickly as possible.

In the world of naval operations, ships are usually given three types of assignment that might constrain their freedom of maneuver: operating within a geographic boundary, transiting from point to point, and stationing relative to another ship.

Operating Box

Operating within a boundary, often referred to as remaining in an operating box, provides the greatest shiphandling flexibility, because the conning officer can choose from any of the 360 degrees to find the best ride. If the goal is to keep the seas on the bow and stern, running reciprocal courses across the operating box allows the shiphandler to keep the seas in a favorable position for the vast majority of time in the box. As long as the turns are executed smartly at increased speed, the seas are unfavorable only for brief periods of time.

The ship may be tasked to patrol the entire box, but following reciprocal courses does not necessarily mean following the same path back and forth. As in a parallel pattern for search-and-rescue operations, turning to the same geographic side at the end of each leg—for example, when following legs of 270 and 090 degrees true, always turning to the north side of the track—will offset successive legs from the previous legs. Using a thirty-degree waterjet angle will offset each leg by the ship's reduced tactical diameter—about a thousand yards, or a half-mile. The pattern can then be reversed when reaching the north side of the box by making every turn to the south side of the track. As illustrated in figure 8-1, running this pattern will ensure that the entire operating box is covered.

Point-to-Point Transit

When transiting from one point to another, the novice shiphandler often drives the ship like a train following a railroad track, but the ship's track is not fixed in the water. It is simply a safe path delineated on a nautical chart. The leading quartermaster will lay the track, and then the navigator will review it for any nearby

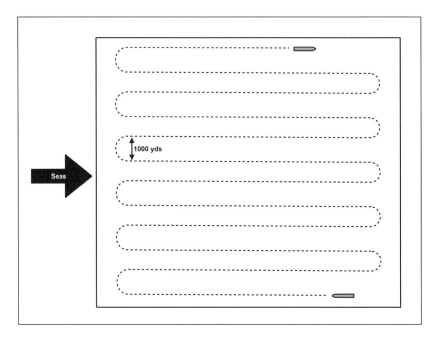

FIGURE 8-1   Patrolling a Box with Two Courses

hazards along the way and ensure that the transit speed, or speed of advance, allows the ship to meet its next commitment. Finally, the track is reviewed by the executive officer and approved by the commanding officer. These layers of checks and rechecks, culminating in the captain's authorization, can be intimidating to a young shiphandler, who may not feel senior enough to question the oversight applied to the track. In most cases, however, the track is laid without considering the direction of the seas. The captain may have approved the track for safe navigation, but the shiphandling team must keep the ship and crew safe through deliberate seakeeping.

This expectation does not imply that the shiphandler is empowered to change the approved track unilaterally, but the captain's standing orders should provide some leeway in maintaining the plan of intended movement (PIM), which is composed of courses and speeds. For example, the captain may allow the shiphandling team to stray five miles laterally from the track in open ocean and up to four hours ahead or behind PIM. When the navigation team reviews the track and when the captain approves it, they are well aware of the leeway authorized in the standing orders. They are looking for hazards not just under the plotted track but also to the left and right side as far out as the shiphandler is authorized to roam. This maneuvering room allows the shiphandler to envision PIM not as a railroad track but rather as a moving rectangle. If the transit is planned for fifteen knots and the standing orders permit deviating from PIM up to five miles and four hours, the shiphandler can maneuver anywhere within a box measuring ten miles by 120 miles.[4] Viewed this way, the track can be considered an operating box that moves at PIM speed, a concept that provides more than enough maneuvering room for the shiphandler to control the ship's movement without straying too far from the plotted track. The novice shiphandler should note, of course, that even when driving within the authority of the standing orders, it is always smart to keep the captain informed of your shiphandling plan.

This approach to open-ocean shiphandling empowers the conning officer to pursue proactive seakeeping even when committed to an approved track; when the seas present on the beam, the conning officer can tackle the issue with a problem-solving approach rather than consider it a situation to be endured. Consider a transit where the ship is following a track plotted along 270 degrees true at fifteen knots. When turning onto this leg, the conning officer finds that the seas are on the beam from 180 degrees true and produce rolls that may hazard the ship and crew. Knowing that seas on the quarter will provide a relatively better ride than on the beam, the conning officer comes right forty-five degrees to a course of 315 degrees true and increases speed five knots; these actions will stabilize the ship. The conning officer can hold this course until the ship is laterally separated from the plotted track by five miles and then turn back toward the track. Altering course to port ninety degrees and steering 225 degrees true places the seas broad on the bow, at 315 degrees relative. The conning officer can continue on this course twice as long as on the previous leg, because the ship can remain on it until the lateral separation is five miles on the other side of the track. This approach will seem familiar to any shiphandler with sailing experience, since it closely resembles tacking for a vessel under sail.[5] The conning officer can continue tacking backing and forth across the track, keeping the seas away from the beam while maintaining PIM.

Chapter Three argued that there is diminishing value on precise math calculations in shiphandling, because the ship is operating in a dynamic ocean environment where the uncontrollable forces generally foil the attempt at precision. The math behind the tacking method is worth exploring here, however, to consider how to manage tacking courses that add distance traveled relative to PIM, because the math reveals the range of options available to the shiphandler. Consider the same scenario, but in this instance the seas are so rough that placing them forty-five degrees off the beam on the first leg is not enough to control rolling. Understanding that moving the seas closer to the stern will reduce roll, the conning officer can alter course to starboard in five-degree increments until satisfied with the ship's movement, but each time the ship steers away from the base course, time is lost relative to PIM. Table 8-1 lists the possible courses and the speeds required to maintain PIM. In the above example, steering a course of 315 degrees true at twenty knots would cause the ship to fall behind PIM one minute for every leg, so the conning officer must increase speed.[6] Table 8-1 shows that twenty-one knots would be required to maintain the fifteen-knot speed of advance along the plotted track.[7]

Figure 8-2 illustrates how tacking between 320 and 220 degrees true at twenty-three knots will maintain the ship's PIM of 270 degrees true at fifteen knots; table 8-1 details the range of options available to the shiphandler. In particular, notice that the range of courses, given the ship's maximum speed of forty-five knots, extends all the way to 340 and 200 degrees true. This course would place the seas twenty degrees off the stern and bow, in a much more advantageous relative position. Altering course by seventy degrees might otherwise seem impossible, but the high speed of which this ship class is capable opens up many more possibilities.

TABLE **8-1** Tacking Courses and Speeds
(Given a base course of 270 degrees true at fifteen knots)

STARBOARD TACK (DEGREES TRUE)	PORT TACK (DEGREES TRUE)	SPEED REQUIRED (KNOTS)
315	225	21
320	220	23
325	215	26
330	210	30
335	205	36
340	200	44
345	195	58

Table 8-1 also shows how the shiphandler can determine the limits based on the engineering plant status. If all four engines are available, the conning officer can alter course as far as 340 and 200 degrees true, but any speed-limiting casualties would further limit the range of options. The conning officer would use the maximum speed available to discern the maximum tacking courses possible. For example, if an *Independence*-variant ship had a casualty to one of the gas-turbine engines, the best speed would be about twenty-six knots, so the maximum tacking courses would be 325 and 215 degrees true.

Take this maneuver one step farther to consider how to manage seas on the bow and stern when they are pushing the ship to its structural limits. As previously mentioned, placing the seas on the bow risks bow slamming, and seas on the stern that outpace the ship can cause erratic motion. This choice drives the shiphandler to slow down in head-on seas and speed up in following seas. The speed listed in table 8-1 are in fact average speeds required to maintain the planned speed of advance, so if half of the legs are into the seas and half are with the seas, the conning officer could manage the speed to achieve an average speed across both legs. For example, if the ship is following tacking courses of 315 and 225 degrees true and needs to average twenty-one knots to maintain

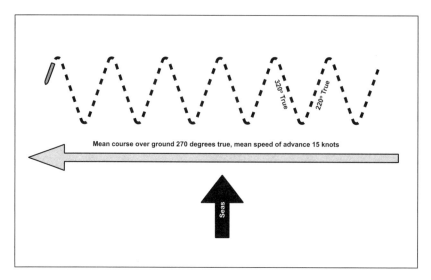

Mean course over ground 270 degrees true, mean speed of advance 15 knots

FIGURE 8-2   Tacking for Seakeeping (Where 320 and 220 Degrees True at
             Twenty-Three Knots Maintains a PIM of 270 Degrees True at
             Fifteen Knots)

the speed of advance, the conning officer could order twenty-six knots when running with the seas and sixteen knots with head-on seas. Over time, this approach would produce an average of twenty-one knots.

Working through the mathematics is useful for understanding the principles, but unless the shiphandler has the time to work through trigonometry at sea, the math is impractical to the shiphandler proactively seakeeping in rough weather. Consider how the tacking method would be used in real time. Heading down the plotted track of 270 degrees true at fifteen knots with the seas on the beam from 180 degrees true, the conning officer would note the undesirable rolling and alter course to 315 degrees true to place the seas on the port quarter. Knowing that the ship is traveling a greater total distance than the plotted track, the conning officer would increase speed to twenty knots. The conning officer may notice after several legs that the ship is slowing falling behind PIM and so increases speed to twenty-four knots. Over time, if the ship regains PIM and begins to spring ahead, the conning officer would know that the ship is moving too fast. Employing the by-half technique described in Chapter Five he or she could reduce the speed to twenty-two knots. If the ship is still gaining on PIM but at a slower rate, speed could be reduced again by half, to twenty-one knots.[8] In just a few iterations, the conning officer will find the ideal speed to maintain PIM—without requiring a scientific calculator—in another demonstration of the continuing balance of the art and science of shiphandling.

The above discussion illustrates how the shiphandler can adjust speed to remain on PIM while tacking, but falling behind PIM is not a situation that must always be avoided. For example, if carefully managing fuel-burn rates, it may be preferable to limit the ship's speed during long transits. In the above case of following a track at 270 degrees true and fifteen knots, the conning officer, after tacking to starboard by forty-five degrees and increasing speed to twenty knots, would discover that the ship was falling behind PIM about one minute for each stretch that the ship traveled from the plotted track to the outermost edge of the five-mile limit. This speed deficit is not especially significant from a PIM-management perspective. If the ship began tacking while exactly on PIM, it would remain within four hours of PIM for the next 1,200 miles along the plotted track. At twenty knots on this tacking course, the ship is making 14.3 knots along the plotted track, so it would take about three and a half days to break PIM limits. Even if the ship were on a course for this long, which is possible on a transoceanic route, it is unlikely that the seas would remain on the same bearing for this amount of time. The shiphandler can return to courses that regain the lost time once the seas shift.

Ship Stationing

In the final example of how ships are constrained in their course—stationing relative to another ship—the conning officer is not following a plotted track. Instead, station keeping requires that the ship maintain a certain bearing and range from the guide ship;

the responsibility for navigating the formation rests on the guide's shiphandling team.[9] This arrangement can be problematic for the ship taking station on larger vessels, such as aircraft carriers; seas that cause excessive movement on this ship class will hardly budge a carrier, so the guide's conning officer will not likely manage courses with seakeeping in mind.

This mismatch does not always compel the shiphandler to follow the guide's course blindly and endure the rough seas. In screen formations, ships are generally ordered to sector stations that provide the flexibility to pursue a multitude of courses. The base course and speed can be treated as relative from a point centered in the assigned sector at a fixed bearing and range from the guide. Instead of bounding the tacking courses by the maximum allowable cross-track distance outlined in the standing orders, the ship's movements are constrained by the limits of the assigned sector. For example, if the guide ship is on a course of 000 degrees true at fifteen knots and own ship is assigned a sector of 240–290 degrees true from the guide at a range of between two and eight miles, the conning officer could envision the same moving box as described above for a plotted track, as illustrated in figure 8-3. This time, the rectangle is oriented to 000 degrees true, centered 270 degrees true and five miles from the guide and two miles across. Instead of tracking distance ahead or behind of PIM, the conning officer would compare the ship's position to a line drawn 270 degrees true from the guide. Viewing ship stationing through this lens and proactively managing seakeeping through the tacking method will provide the shiphandler a range of options similar to those in table 8-1 above.

Of final note, even if the bridge team on the carrier does not fully appreciate the seakeeping challenges of this ship class, it should respect feedback with respect to station assignments. If the assigned stationing sector does not provide enough freedom of maneuver to manage seakeeping adequately, ask for an expanded sector. Professional mariners know the limits of their ship and will not endanger the ship or crew out of pride.

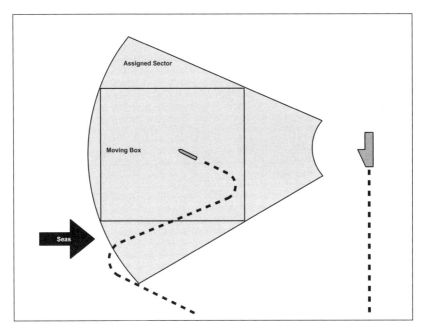

Figure 8-3   Tacking in an Assigned Sector

THE ROLES OF CHARTS AND FORECASTING IN FINDING SAFE WATER

In addition to managing seakeeping by positioning seas relative to the ship, proactive seakeeping includes reviewing weather forecasts to locate good seas. Wave heights can change by many feet over just a few dozen miles. This is particularly true for the nearshore environment, where this ship class frequently operates.

Topography considerably impacts wave heights. High winddriven seas develop in areas of unlimited fetch, where as noted in an earlier chapter the wind travels across hundreds of miles of water without disruption. Once wind encounters land, such as an island or peninsula, the seas on the leeward side—in the shadow zone—will have waves far less severe than on the windward side. Locating these shadow zones is fairly simple even without modern forecasting. For example, consider an island with the wind coming from 315 degrees true, as illustrated in figure 8-4. On the chart,

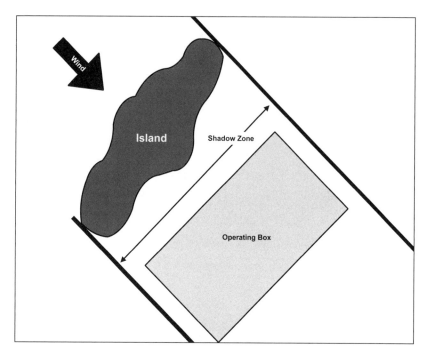

FIGURE 8-4   Operating in the Lee

the shiphandler can draw from the northeast and southwest ends of the island lines reciprocal to the wind direction—that is, 135 degrees true and tangent to the island. These lines can be used to bound a rectangular operating box on the northeast and southwest sides, with perpendicular lines lying 225 to 045 degrees true on the northwest and southeast sides. The size of the box can be expanded by moving the southeast boundary farther from the island, but the farther the box extends from the island, the more likely that the seas will refract around the island and produce larger wave heights.

Additionally, underwater topography affects what types of seas are experienced on the surface. As a rule of thumb, the hundred-fathom curve offers a good marker for expected seas on the seaward and landward side of this line. On the seaward side, sea swells perpetuate without obstruction, often yielding severe, long-period swells that are independent of seas driven by local wind. Seas with relatively low significant wave heights (meaning average height of the one-third highest waves) may still generate swells that produce considerable heave, affecting this lightweight ship. Longer swells are preferable for large, deep-draft ships, which barely feel the heave, but they are quite noticeable for three-thousand-ton vessels. On the landward side of this curve, the rising sea floor begins to erode this energy, lessening the heave produced and making the swells more manageable. Long-period swells yield to seas driven by local winds, seas that are easier both to predict and manage.

Beyond studying topography, the shiphandler should regularly consult modern forecasting reports to locate the best water for seakeeping. National Oceanic and Atmospheric Administration (NOAA) buoys provide significant-wave-height measurements every hour, and NOAA forecasts provide expected wave heights out to seventy-two hours with reliability. The seakeeping challenge is best tackled with a balance of the art and science of shiphandling. Scientific forecasting tools allow the shiphandling team to select the most advantageous operating areas for the ship; once in position, the conning officer can apply the art of shiphandling to make the best of the uncontrollable forces presented.

Whether its crew uses charts or forecasts, this ship class possesses a key attribute for seeking the best water for seakeeping. The ship's high speed allows it to reposition quickly, so rapidly deteriorating seas can be escaped when necessary. If good water lies sixty miles away, a propeller-driven ship may think that the benefit is not worth a three-or-four-hour transit. This ship class's speed makes that area reachable within ninety minutes, so the ship could sprint to the new station with relative ease. When managing issues like fatigue, these conning officers are uniquely empowered to improve the safety and readiness of the crew within minutes of making the decision to do so.

CONCLUSION

This chapter opened by promising to discuss methods for seakeeping when operational commitments seem to conflict with courses that would provide for the best ride—and the critical qualifier in that promise is "seem." The novice shiphandler often looks at an operational commitment and assumes more constraints than really exist. In the end, the ship must arrive at a certain location or remain within a certain area—moving or stationary—but how it accomplishes the goal is largely up to the shiphandling team. Sea-keeping is merely another problem-solving exercise, requiring the shiphandler to balance competing requirements to find an optimal solution that accomplishes both operational and seakeeping objectives. While meeting operational commitments will always serve the ship well, doing so at the expense of the sailors indicates a shiphandler who is inexperienced, shortsighted, or obtuse.

The tools presented here allow the shiphandler to accomplish both objectives with a modicum of forethought, followed by the active shiphandling management that would be expected during any watch at sea. Only in the rarest circumstances will the cost-benefit calculus indicate that good seakeeping disadvantages the ship; in most cases, the shiphandler should instinctively lean toward a plan that provides for a rested and productive crew.

CHAPTER NINE

Simulator Training

One of the best opportunities for shiphandlers in this ship class is the extraordinary amount of training conducted ashore during the off-hull period, a time when the crew works from office spaces and prepares to go back on board for the next on-hull period. These ships are manned by rotational crews, with time ashore dedicated to training in all aspects of shipboard duty—combat operations, damage control, engineering operations, first aid, security, navigation, seamanship, and shiphandling. The culminating event for the training season occurs in the simulator, in which the crew is prepared and tested on shiphandling, navigation, and combat operations through hundreds of hours in the virtual trainer.

A typical day in the trainer includes navigation and combat training for each of the ship's watch teams, training that runs from early-morning until late-afternoon, all under the guidance of a professional training staff. Upon completion of the structured events each day, the commanding officer is afforded time into the evening to work individually with the shiphandlers to hone their skills prior to driving the actual ship. The commanding officer can tailor this unstructured shiphandling training time to specific upcoming events or focus on helping novice shiphandlers. Without a plan for training, however, this unstructured time can become a wasted opportunity.

The philosophy of training in simulators for this ship class is similar to that of simulator training, which has been in the fleet for many years. By offering a risk-free environment in which to practice shiphandling skills, trainers allow novice shiphandlers to learn from mistakes that would not be permitted at sea. This is not to say that conning officers do not learn from mistakes at sea, but rather that in a simulated environment the conning officer can take the ship all the way to the point of collision. At sea, the commanding officer would step in long before the ship reached the point of extremis, let alone collision. This invaluable aspect of training will go unrealized, however, if mistakes in the trainer do not translate into lessons for the shiphandler.

This chapter aims to provide advice on how to maximize this valuable opportunity to train young shiphandlers, beginning with a discussion about the role of the advanced shiphandlers in the training process. Then, it will present a ten-day training plan designed to guide novice shiphandlers through exercises of increasing difficulty and give them the confidence to tackle close-quarters maneuvering once back onboard the ship. This chapter will conclude by offering a small set of skill drills that can be practiced whenever the shiphandling team has a few extra minutes of unstructured time in the trainer.[1]

ROLE OF THE ADVANCED SHIPHANDLERS

Young officers make for great shiphandling students, largely because they are eager to learn and excited at the prospect of driving a real warship. At a relatively young age, they are given the responsibility of controlling a sizable and complicated national asset, responsibility that would seem unfathomable to peers outside their profession. This has always been true in naval shiphandling, but this ship class provides an even more exciting aspect for these aspiring ship drivers, in that they have their hands physically on the ship's controls. They are not merely ordering someone to drive the ship but are driving the ship themselves.

If youthful eagerness is their best attribute, their greatest shortfall is a lack of experience and perspective in the art and science of shiphandling. How fast is too fast? How slow is too slow? How close is too close? How much power is too much power? The experience and perspective gained from years of shiphandling at sea is less about real values than it is about relative values, and these relative values are largely situational. Transiting at thirty knots in the open ocean is different than thirty knots in a narrow channel. Making a five-knot approach toward an anchorage is different than a five-knot approach toward a quay wall. Novice shiphandlers, whose relative motion experience relates more closely to cars than ships, may not see the danger in landing the ship at five knots. After all, five knots is less than six miles per hour, and driving a car at six miles per hour seems conservative. The experienced shiphandler has developed a calibrated sense for speed and momentum that is difficult to codify, but when the ship is moving too quickly to be safe, an uneasy feeling prompts a quick reduction in speed.

Honed over many years of at-sea shiphandling, this perspective is invaluable in the simulator. Most novice shiphandlers are receptive to the time-tested principles of shiphandling, having no experience base against which to compare their performance, and they are seeking feedback that will help them understand if they are doing it correctly and how they can do it better. In the example above, if the novice barrels in toward the quay wall at five knots, the experienced shiphandler can lean in and simply say, "Five knots is too fast for this maneuver." Such simple feedback is possible only from someone who has seen countless docking evolutions.

Furthermore, the experienced shiphandler can explain why a simulator landing has just turned into an allision. Young officers left to themselves in the simulator may allide with the pier, hear the computerized crashing sound, infer, "Well, that wasn't right," and leave it at that. Self-assessment has limited effect when the conning officer understands only one or two of the factors involved. For example, after learning that too much speed in the basin causes allisions, the novice may conclude that every allision is caused by excessive speed, when any of the controllable or uncontrollable forces can be the actual cause. The experienced shiphandler can stand back, take into account all the factors, and provide feedback immediately after the allision.

Finally, experienced shiphandlers in the simulator introduce realism into an environment that often deteriorates into an unproductive video game when unsupervised. Certain shiphandling behavior would never be tolerated at sea, so there is no point in practicing it in the simulator. For example, curiosity might lead one to wonder whether the ship can enter the basin at twenty knots and stop in time with a full backing engine, but there is little enduring value in practicing this maneuver. Experienced shiphandlers in the simulator should encourage young conning officers to practice the simple, repeatable methods that will make them successful shiphandlers at sea.

It is important to note, however, that the trainer cannot entirely simulate real shiphandling, because driving the ship involves the real risk of damaging the real ship. The trainer provides a safe environment in which the shiphandler can practice without consequences, but the pressure of those possible consequences will produce a fundamental change in the shiphandler. Stress causes physiological changes that affect decision making. The presence of the commanding officer and executive officer in the trainer is critical to enhancing the realism of simulator training. A young conning officer would never be left alone to dock the ship in a harbor; the commanding officer will likely be within arm's reach, and the executive officer not much farther. Recreating this arrangement in the simulator conditions the novice shiphandler to the pressure of driving while standing next to the captain. Furthermore, if the ship's two most experienced shiphandlers foster an environment where professionalism is displayed by pride in practicing one's profession, every simulated docking and undocking will become an opportunity for the young officers to demonstrate their shiphandling skills to the captain.

THE TEN-DAY TRAINING PLAN

When developing a training plan, a wise approach is to increase difficulty progressively to allow the novice shiphandlers to develop skills. The idea in progressive shiphandling training is to begin with simple maneuvers under benign conditions, building to simple maneuvers under difficult conditions, followed by complex maneuvers under benign conditions, and finally complex maneuvers under difficult conditions. In the sample schedule below, consider how shiphandling evolutions develop over weeks. This plan assumes new shiphandlers who have just reported on board, who have had basic shiphandling training for this ship class but need to solidify their skills. In addition, given the rotational crewing plan, even shiphandlers with onboard experience will likely require refresher training to regain their shiphandling acumen after several months ashore.

Day One

The goal of Day One is to exercise the controllable forces. To focus the shiphandlers on this skill, the simulator is set to omit the effects of uncontrollable forces. Winds and current are both set to zero. Additionally, basin obstructions are eliminated by clearing the pier and adjacent piers of ships, so that the conning officer needs only to push away from the pier and drive into the channel. Once the ship is clear of the basin, the conning officer twists the ship and then docks in the same basin on the adjacent pier, as illustrated in figure 9-1.

In this fairly basic maneuver the conning officer would, as discussed in Chapter Five, on pierwork, toe-out both engines (position A), with the starboard engine ahead, port engine back, and thruster trained to 270 degrees relative. The ship can also be walked laterally without a thruster under these environmental conditions, or the conning officer can practice having the tug pull to port. Once the ship is clear of the pier hazard (position B), the shiphandler would increase thrust on the starboard ahead engine to build headway and when in the channel commence a twist to port. The twist (position C) would involve canceling longitudinal vectors, toeing-in both waterjets, and increasing power to the thruster (or tug). As the ship points into the basin, the conning officer would center the waterjets, stop the thruster (or tug), and increase thrust

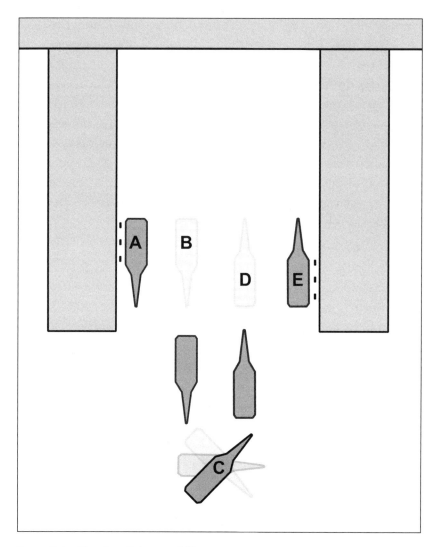

FIGURE 9-1   Day-One Scheme of Maneuver

on the starboard ahead engine to build headway. Headway would be stopped when the ship sits adjacent to the new berth (position D) by checking forward motion with increased thrust on the port backing engine. To walk laterally to starboard, the conning officer would toe-in both engines and train the thruster to 090 degrees relative (or have the tug push to starboard) until alongside the pier (position E)—but again, the thruster and tug are not required for this maneuver. The details for controlling the ship in terms of the three gauges for pierwork—heading, longitudinal speed, and lateral closure—have been discussed in detail in Chapter Five. This sequence allows the shiphandler to exercise total control over the ship with a straightforward undocking and docking evolution in the same basin and reinforces the concept that the stern can be moved laterally in either direction without changing the thrust direction of the engines. In this maneuver, the ship walks off from the pier in a toed-out configuration and walks onto the pier in a toed-in configuration.

Day Two

On the second day the previous day's evolution is repeated, but this time the shiphandler must manage wind. Still, the addition of an uncontrollable force poses only a modest challenge, with the wind steady at five knots, onsetting for the undocking and offsetting for the docking. This wind arrangement is used as the first introduction of uncontrollable forces for several reasons. First, the ship is capable of easily walking against a ten-knot wind without a thruster or tug. Second, a steady wind presents a simpler shiphandling challenge than does a gusting wind, which forces the conning officer to adjust constantly the application of lateral force to account for varying winds. Finally, it is easier to land against an offsetting wind than to brake against an onsetting wind pushing the ship into the pier.

The maneuver to be executed will closely resemble that of Day One, but the conning officer must adjust to the added force acting on the ship. With an onsetting wind during the undocking, the shiphandler will use more lateral force with the waterjets and thruster (or tug) to break the ship from the pier, though with the wind under ten knots the ship can still be laterally walked without a thruster. The most valuable part of this exercise is the twist. The wind will move the ship back across the slip as it twists, and the conning officer must apply enough lateral force to hold the ship up into the wind to prevent it from blowing back into the pier it just left. Once back in the slip, the shiphandler will again need to use more lateral force, with the waterjets and thruster (or tug), to dock the ship against the offsetting wind.

Day Three

After this maneuver has been accomplished with the wind onsetting during undocking and then offsetting during docking, the wind direction is reversed to present more challenging conditions. The wind is still set at a steady five knots to allow the shiphandler to maneuver without a thruster or tug if desired. By the third day, it may be tempting to change the pier arrangement, but controlling the number of variables from day to day accomplishes two goals. First, it gives novice shiphandlers confidence that they have already accomplished a similar maneuver; second, it focuses them on the true task for the day—in this case, docking the ship with an onsetting wind.

The offsetting wind for the undocking will make it easy to get off the pier. Once twisting in the channel, the task of holding the ship up into the wind will be similar, but in this case the conning officer must keep the ship laterally in place or risk being blown across the slip and into the opposite pier. Walking the ship laterally toward the pier once in the slip will require little power to build momentum; with the wind creating a discernible closure rate toward the pier, it may not require any lateral controllable force at all. While the novice shiphandler will want to point the controllable forces toward the berth, the advanced shiphandler knows to keep the controllable forces pointed into the wind as a braking mechanism. On Day Three, the shiphandler learns to allow Mother Nature to dock the ship and to use the ship's power to ease the landing.

Day Four

The previous two days of practicing shiphandling in differing wind conditions have taught the novice shiphandler to apply controllable forces to both counter and leverage a single uncontrollable force. On Day Four a second uncontrollable force, current, is added to make the conning officer determine which uncontrollable force is causing which ship movement. For this reason, it is most effective to apply a current that runs counter to the wind. With the wind configuration the same as on Day Three, the current would be set to run from downwind to upwind at about one knot. By the rule of thumb for wind and current effects for this ship class discussed in Chapter One—fifteen knots of wind is equal in force to one knot of current—the current would have slightly more effect than the wind.

The maneuver off the pier is similar to Day Three, when the offsetting wind allowed the ship to break away from the pier using very little power. The solid pier will largely block the effects of the current, so the new challenge presents itself when the ship develops headway and the bow protrudes into the channel. The wind blows from starboard to port, and the current runs from port to starboard, so moving the bow into the channel while the stern is still in the slip exposes the ship to two counteracting forces. The wind will continue to push the stern to port, and the current, the more powerful force, will push the bow to starboard, resulting in a twist to starboard. The stern will swing across the slip, but the bow will move toward the pier hazard and adjacent buoys. The conning officer must recognize the effect of the current and use the thruster (or tug) to hold the bow up into the current while continuing to hold the stern up into the wind. This specific moment in the exercise will highlight the importance of understanding the six shiphandling actions described in Chapter Three, on waterjet vector management. When the ship is twisting in the channel the conning officer must apply enough lateral force to keep the current from carrying the ship away from the slip entrance. Once the ship begins moving back into the slip, the challenge of managing opposing uncontrollable forces presents itself again, with the bow finding shelter from the current while the stern still feels its effect. The ship will again fall solely under the wind's effects once completely inside the slip, and the conning officer must dock the ship with an onsetting wind.

Day Five

By the end of Day Four, the shiphandlers should be well practiced at undocking and docking on a pier with various environmental conditions. Day Five presents a more challenging maneuver for the conning officer. Undocking from a quay wall situated between two piers, the conning officer must walk off the quay wall, pivot the ship on the bow, and then back straight out of the slip, as illustrated in figure 9-2. Both piers can be lined with docked ships to create an increased level of difficulty. Once the ship has exited the slip and entered the channel, the conning officer will stop the stern-way, build headway to reenter the slip, and then dock the ship in the original berth by pivoting on the bow. As on Day Two, this

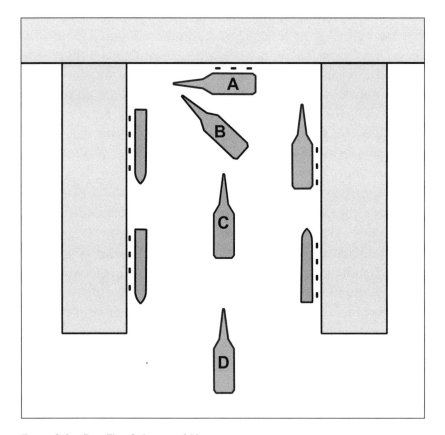

FIGURE 9-2   Day-Five Scheme of Maneuver

challenging maneuver is best attempted for the first time with an offsetting wind to aid the undocking and buffer the landing.

With the ship berthed starboard side toward the pier (position A), the conning officer begins by walking laterally to port, with both waterjets toed-out, starboard engine ahead and port engine back, and the bow thruster trained to 270 degrees relative (or tug pulling to port). Preventing longitudinal movement becomes critically important, as the piers both ahead and astern limit the ship's freedom of maneuver forward and aft. Once the conning officer has created sufficient room between the ship and the quay wall, the pivot maneuver may commence. The pivot point is moved all the way to the bow by increasing the toe angle while increasing power on the bow thruster (or tug) just enough to prevent the bow from twisting to starboard (position B). This technique of anchoring the bow allows the stern to swing about to port and into the center of the slip. As discussed in Chapter Five, the conning officer on an *Independence*-variant ship should practice dynamic thruster management, systematically adjusting the thruster away from the pier hazard relative to the ship's heading. Training the thruster aft of 270 degrees relative will move the pivot point farther aft and twist the bow into the center of the slip. When the ship is centered between the piers, the thruster should be pointed 180 degrees relative.

Once the ship is centered (position C), the conning officer develops sternway by increasing power on the port backing engine. Adjusting the waterjets using the toe-in/toe-out method will allow the conning officer to keep the stern centered in the slip, toeing-out to move the stern to port and toeing-in to move it to starboard. The thruster (or tug) can be employed as necessary to center the bow as well. The conning officer backs out until the ship is in the channel (position D) and then stops the ship's sternway and builds headway by easing off the port backing engine and increasing power on the starboard ahead engine. The thruster, if in use, should be pointed back into the slip, and the conning officer again exercises the toe-in/toe-out method to hold the stern in the center of the slip while approaching the quay wall (position C).

As the ship closes the quay wall, the shiphandler aims the bow into the corner between the pier and quay wall, by either pointing the thruster toward the corner or pulling to port with the tug. This lateral force will cause the stern to pivot to starboard, so if ships must be passed down the starboard side on the adjacent pier, the conning officer toes-out both waterjets to hold the stern in the center. Once the stern is clear of obstructions, the conning officer toes-in the waterjets to move the stern toward the quay wall. This move will cause the bow to twist to port, so anchoring the bow will keep it adjacent to the quay wall as the stern swings to starboard (position B). The bow is held in place by training the thruster to 090 degrees relative (or pushing ahead to starboard on the tug) and applying just enough power to hold the bow's closure rate toward the pier at zero. When the ship is parallel to the quay wall, the conning officer transitions from a pivot on the bow to a lateral walk by increasing power on the bow thruster (or tug) and reducing the toe angle to slow the stern's closure rate toward the pier (position A).

This series of maneuvers exercises many advanced shiphandling skills. The shiphandler must walk the ship laterally with a only small margin for longitudinal movement and then actively move the pivot point from amidships to the bow to swing the stern out, all while employing dynamic thruster management. When backing out of the slip, the conning officer uses the toe-in/toe-out method to keep the stern centered while maintaining sternway. Stopping the ship in the channel and moving ahead into the slip again practices longitudinal control over the ship, and pivoting back toward the quay wall in a toed-in configuration reinforces the idea that the stern can be moved laterally in either direction without changing the thrust direction of the engines. In this case, the ship walks off the quay wall in a toed-out configuration and walks back onto the quay wall in a toed-in configuration.

Day Six

Much like the progression from Day Two to Day Three, the wind direction is reversed on Day Six to present a more challenging docking situation. The same quay wall maneuver is repeated, except this time the wind is onsetting. These conditions mean that more power is required to push off the quay wall, and during the docking maneuver the onsetting wind will push the ship toward the hazard, requiring the conning officer to use controllable forces to brake against the wind's effects. For an added level of difficulty, the wind can be set to 240 degrees true, given that the ship is pointing 270 degrees true when alongside the quay wall. This angle will provide a wind that is onsetting when near the quay wall but crossing when moving in the slip, a condition that challenges the conning officer to keep the ship centered.

Day Seven

The progressive approach to practicing pierwork culminates in a repetition of this quay wall docking and undocking but with even more challenging uncontrollable forces. In addition to the onsetting wind, arranged to blow across the slip when the ship is backing out, as described for Day Six, this maneuver includes a channel current that moves from starboard to port. Additionally, the wind is set to gust up to twice its sustained speed. For example, in a sustained wind of ten knots, it would occasionally gust up to twenty knots.

The introduction of current to this exercise is analogous to the transition from Day Three to Day Four. The ship will not experience much effect from the current while in the slip, but once the stern emerges from protection the current will push it to port. The conning officer must toe-in the waterjets to hold the stern up into the current, and it will be necessary to use the thruster or tug to keep the bow from falling into the ship berthed on the adjacent pier. The bow will feel this same effect once the entire ship is in the channel, so the conning officer will need then to use controllable forces on both the bow and stern to hold the ship up into the current. Reentering the slip, of course, will be a reversal of this sequence.

The most difficult challenge in this exercise, however, is the introduction of gusting winds, particularly while docking the ship.

Shiphandling is always more difficult when winds vary in speed, but strong, gusting winds present a particular problem in pierwork. Wind gusts require a quick application of power to keep the ship from blowing into a hazard, but as soon as lateral power is applied the gusts might fall off, leaving the ship applying power against an uncontrollable force that is no longer there. Leaving this lateral force on too long will result in an overcorrection. Reversing the lateral force again to counteract this overcorrection has its own risks, as the effects of the gusting wind would be compounded if the controllable forces are pointed in the same direction as the wind. Needless to say, combating a gusting wind requires an awareness of relative motion, a sense for wind speed changes, and a well-trained shiphandling team that can nimbly adjust controllable forces to offset the uncontrollable forces.

Day Eight

Pierwork is the most challenging shiphandling exercise, so pierwork practice consumes seven of the ten days; once Day Seven concludes, the conning officers can proceed to other evolutions. Anchoring is fairly straightforward, but novice shiphandlers must learn to control the ship's momentum. As described in Chapter Seven, on special evolutions, appreciating relative motion in this regard is difficult at first, because there are no markers in the anchorage area to make the ship's relative speed apparent. Five knots when approaching an unmarked anchorage seems negligible, especially when compared to how five knots appears bearing down on a quay wall. The shiphandler must learn to develop a sense of momentum from the ship's gauges. Through trial and error in the simulator, the novice will learn how much power is required to slow the ship to a stop and then develop slight sternway.

During the anchoring, the conning officer should approach the VMS-plotted anchorage from a distance of one mile at fifteen knots. Following a plotted head bearing is not vitally important, as described in Chapter Seven, as long as VMS is checked for shoal water and underwater obstructions. The more important skill is considering the uncontrollable forces that will push the ship around as it approaches the anchorage. For example, in a protected harbor the conning officer should make the approach into the wind or current, whichever is the prevailing uncontrollable force. In an unprotected anchorage, comparing the wind and seas will help determine the approach course.

When the approach commences, the conning officer should focus on steering a course for the anchorage and controlling the ship's speed. It is important to remember that steering directly toward the anchorage becomes less significant with each yard closer the ship comes to the anchorage. The most critical gauge is the bearing drift. If the center of the anchorage is drifting to the left, the conning officer must steer left of the bearing to the anchorage. Conversely, if the anchorage is drifting to the right, the ship will require a course right of the bearing to the anchorage. The conning officer will know that the selected course is correct when the bearing remains constant. Given that the ship is approaching at fifteen knots, the conning officer should reduce speed to ten knots at a thousand yards, then to five knots at five hundred yards. When the ship is one hundred yards from the anchorage, backing down on both engines will stop the ship's headway, and the anchor should be released when sternway develops. Of note, the simulator does not permit the conning officer to discern sternway by watching objects in the water, so GPS becomes the primary tool for verifying the ship's speed.

Day Nine

Modern simulators are valuable for every aspect of shiphandling but especially for underway replenishment, a particularly hazardous evolution that is made safer by repetition in the trainer. In the simulator, shiphandlers can take extra time to find the correct lineup, and if their approaches do not work out, they can haul around and try again. Actual replenishment at sea often does not afford this flexibility, as oilers have schedules and other ships are waiting to refuel next. Allowing novice shiphandlers to miss the approach too wide costs time for both the oiler and the ship, and allowing them to miss too close can have more dire consequences. The simulator provides a safe environment for error and helps make novice shiphandlers ready for at-sea replenishment.

When conducting this evolution in the simulator, young conning officers are tempted to rush alongside the oiler, often before they have verified the lineup. As discussed in Chapter Seven, they should take as much time as necessary in standby station to watch two critical factors: relative speed and lateral separation. Verifying that the ship has matched the oiler's speed, by its very nature, takes time; the conning officer must take several range readings to ensure that the ship is neither opening nor closing longitudinally. Similarly, after establishing the parallel index line on the radar display, the conning officer must observe lateral separation over time to ensure that the ship is neither opening nor closing laterally. The simulator provides an ideal opportunity for the novice shiphandler to learn that time spent refining the course and speed in standby station will pay dividends when alongside.

Chapter Seven discussed several methods for making the approach, and again, the simulator is ideal for attempting each method and watching how the ship performs. Whether the conning officer is making an approach for the first time or is an advanced shiphandler honing replenishment skills, mastering different methods for the approach will give them a diverse tool kit that can be applied in varying situations, with options ranging from the coast-in method to the braking method. A strong shiphandling team will be able to apply the right tool to any given situation.

Once the ship is alongside and steady, the novice shiphandler will be eager to break away and practice the approach again. As in standby station, more experienced shiphandlers must encourage young conning officers to remain alongside and practice station keeping. Approaches and breakaways during replenishment at sea may be the most exciting parts of the evolution, but they constitute only a fraction of the total event. The majority of any replenishment is spent alongside, with the conning officer making small adjustments to course and speed to keep the ship in position. This aspect of replenishment can be tedious, but practicing this phase of refueling builds a mental stamina that is critical for safe shiphandling.

In addition to normal breakaways, a final aspect of replenishment, one that is particularly worth exercising, is the emergency breakaway. Emergencies should be generated in varying combinations, on both the ship and oiler. The shiphandlers should practice breakaways with oiler rudder casualties that turn the oiler toward the ship and others away from it, and engine casualties should cause the oiler to drop back and surge ahead sharply. On the receiving ship, casualties should include waterjets that jam toward and away from the oiler, as well as the loss of one engine or several. Much like the challenge of conning alongside, the novice shiphandler will be eager to break away quickly, but the ship should be held alongside the oiler for the time it would take the rig team to take action to prevent injury to personnel or damage to the receiving station. Practicing this skill will be very important in the event of an alongside casualty.

Day Ten

On the final day of this ten-day training plan, the shiphandlers will practice man-overboard recoveries. This shiphandling training, like the others listed above, should center on the realistic problem of retrieving a sailor who has fallen overboard. The novice ship-handler may be tempted to speed around in a circle and get back to the man as quickly as possible, racing the clock to see who can return the fastest, but as described in Chapter Seven, recovering a man overboard involves more than simply colocating the ship and the man.

When practicing this maneuver, the conning officer should demonstrate an ability to evaluate quickly the uncontrollable forces at play. In particular, as presented in Chapter Seven, it is important to envision how the wind and seas will affect the ship after it slows to less than five knots or stops. Most recoveries employ the small boat, so the shiphandler must consider how to position the boat relative to the man so that the boat can be launched with a sufficient lee and have the shortest possible run toward the man in the water. An ideal approach would have the wind and seas on the starboard bow and the sailor on the port side forward of the bridge. Of course, after moving swiftly into position, the ship must slow down enough to launch the boat.

Shiphandling techniques discussed throughout this book envision most maneuvers as using the controllable forces to offset or leverage the uncontrollable forces so as to move the ship into the intended position. What makes the man overboard maneuver different is that it cannot be meticulously planned like an anchorage or docking evolution. The goal of man-overboard training in the simulator is to help the shiphandler develop an ability to assess quickly the uncontrollable forces and plan an effective approach. For some novice shiphandlers, it may be necessary on the first run to pause the simulator to talk through the problem, but in the end the conning officer must be able to make these assessments and decisions in stride.

SKILL DRILLS

Practicing shiphandling in the simulator is like practicing football on the playing field. On some days the coach schedules a full scrimmage to simulate game conditions, but on other days practice centers on drills aimed at improving individual skills. Each drill is designed to focus on one specific skill, with the expectation that improvement in these individual skills, combined with full scrimmages, will prepare the players to win games. The simulator is useful in the same way to prepare shiphandlers to drive ships at sea. Simulator time can be divided between the full shiphandling evolutions presented above and smaller drills that focus on specific skills. Taking advantage of a little extra time here and there, the following skill drills will help shiphandlers hone the techniques necessary to control the ship.

Driving in Squares

Most shiphandling maneuvers follow a direct path toward an intended position. For example, if the ship needs to move forward and to the right, the shiphandler is more likely to crab in that direction than to move longitudinally ahead before laterally walking to starboard. There is a certain training value, however, derived from dividing this maneuver into two separate legs. Driving in squares—moving the ship at right angles, as illustrated in figure 9-3—requires a greater degree of control over the ship. This shiphandling approach is not widely applicable in the real world, but a conning officer who can drive in squares can move the ship deliberately and effectively in any direction.

Leaning into the Current

This simple drill exercises the ability to control lateral forces. Placing the ship in a channel with a strong current on the beam, the drill involves applying just enough lateral force on the bow and

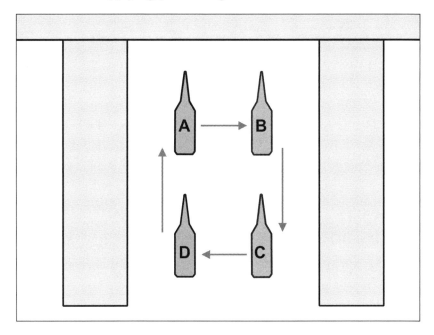

FIGURE 9-3   Driving in Squares

stern to hold the ship in place. Lateral forces on the bow and stern should be pointed into the current; this drill can be done with either a bow thruster or tug. Using the three gauges for pierwork the shiphandler should keep the heading steady and longitudinal and lateral speed both at zero.

Anchoring the Bow

One of the best characteristics of this ship class is that its pivot point can be moved anywhere along the centerline on demand using the controllable forces. Though envisioned largely for the bow thruster on the *Independence* variant, this skill drill is also possible with a tug on the bow. This exercise is termed "anchoring the bow" because the pivot point is moved as far forward as possible, as if a dropped anchor were holding the bow in place. In this drill, the goal is to fix the bow while swinging the stern 360 degrees, as illustrated in figure 9-4.

This maneuver is accomplished by toeing-out the waterjets and applying a counterforce with the bow thruster (or tug) just large enough to keep the bow from pivoting in the opposite direction. For example, the conning officer could begin by toeing-out both waterjets fifteen degrees, with the starboard engine ahead T2 and port engine back T2 and the bow thruster trained to 270 at T5 (or tug pulling to port dead slow). The shiphandler watches the lateral movement of the bow and stern carefully, increasing and decreasing lateral forces as needed to ensure that the stern is moving left and the bow is steady. Throughout the maneuver, waterjet thrust should be adjusted to cancel longitudinal force vectors and thereby produce zero longitudinal speed.

Anchoring the Stern

Like anchoring the bow, this maneuver is accomplished using the bow thruster on the *Independence* variant, although it is also possible with a tug on the bow. The goal in this maneuver is to move the pivot point as far aft as possible; it is termed "anchoring the stern" because the ship behaves as if an anchor had been dropped from

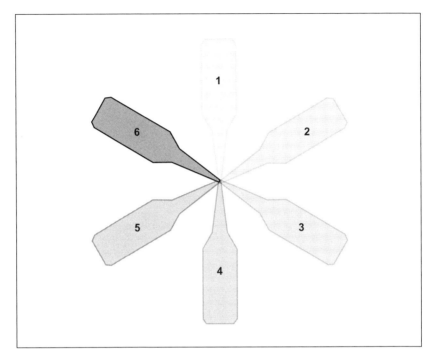

FIGURE 9-4   Anchoring the Bow

the flight deck and was holding the stern in place. As depicted in figure 9-5, the idea is to hold the stern in place while swinging the bow 360 degrees.

This is accomplished by applying maximum thruster power (or moderate tug power) toward either beam, while the waterjets are toed with just enough counterforce to prevent the stern from pivoting in the opposite direction. For example, the conning officer could begin by toeing-in both waterjets ten degrees, with the starboard engine ahead T2 and port engine back T2 and the bow thruster trained to 090 at T10 (or tug pulling to starboard easy). Carefully watching the lateral movement of the bow and stern, the conning officer increases and decreases lateral forces as needed to ensure that the bow is moving right and the stern is steady. As in anchoring the bow, waterjet thrust should be adjusted to cancel longitudinal force vectors to produce zero longitudinal speed.

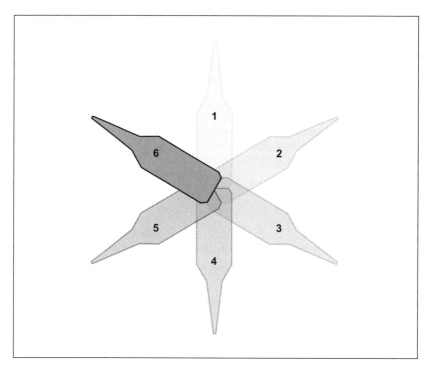

Figure **9-5**   **Anchoring the Stern**

CONCLUSION

This book began by describing the commanding officer's relationship with shiphandling as "intensely personal." As the most experienced shiphandler onboard—with about twenty years of commissioned experience—the captain likely has certain methods or approaches that have served well in the past. He or she would not be in command at sea without a very good record of shiphandling, so it is only to be expected that the methods underlying this record would act as the foundation for command. Moreover, safe shiphandling is an unambiguous performance marker, one that starkly determines whether a command tour is successful or not.

This final chapter may seem highly prescriptive, but a twenty-year veteran of shiphandling understands that not all at-sea experience is the same. For some officers, this training plan may be just the starting point that is needed to develop a tailored program for the ship's team. For others with less experience in developing structured training, it may be the off-the-shelf plan needed to get a team ready. In any case, the principles of modern shiphandling training remain largely separate from the experience of the commanding officer. Training simulators have revolutionized the way in which novice shiphandlers learn the art and science of shiphandling; deliberate planning, with dedicated involvement from the ship's advanced shiphandlers, will ensure that these opportunities are fully realized.

PREFACE

1. R. S. Crenshaw, *Naval Shiphandling*, 4th ed. (Annapolis, Md.: Naval Institute Press, 1975), 335.

CHAPTER 1. INTRODUCTION

1. The word *amah* [ah-mah], originally East Asian, means servant. In this case, the outriggers are performing as servants by assisting the main hull.

2. "Littoral Combat Ship Class: LCS," *United States Navy*, accessed 2 September 2014, http://www.navy.mil/navydata/fact_display.asp?cid=4200&tid=1650&ct=4.

3. To understand the rule of thumb for these ships, consider the difference in surface area between this ship class and the *Arleigh Burke* class. Surface area both above and below the waterline is tied directly to the ship's length; greater ship length increases both surface areas. The critical comparative measure is the draft. The *Arleigh Burke* class has a thirty-foot draft, and as previously mentioned, the *Independence*-variant draft is fourteen feet and that of the *Freedom* variant is thirteen feet. With about half the draft, both variants of the littoral combat ship have roughly half as much surface area beneath the waterline, relative to the ship's length, as the *Arleigh Burke* class. Given that the surface area above the waterline is roughly proportional for the ship's length, we can conclude that the rule of thumb for wind and current can be reduced by about half. Rules of thumb by their very nature are rough estimates, so it is sufficient to presume that one knot of current is equal to fifteen knots of wind.

CHAPTER 2. CONTROLLABLE AND UNCONTROLLABLE FORCES IN SHIPHANDLING

1. Crenshaw, *Naval Shiphandling*, 37.
2. For further discussion of this rule of thumb, see Chapter One, endnote 3.
3. Crenshaw, *Naval Shiphandling*, 38.
4. U.S. Government, *The American Practical Navigator* (Bethesda, Md.: National Imagery and Mapping Agency, 1995), 443.
5. Ibid., 775.
6. Ibid., 443.
7. Crenshaw, *Naval Shiphandling*, 13–15.
8. James Alden Barber, *Naval Shiphandler's Guide* (Annapolis, Md.: Naval Institute Press, 2005), 48.
9. Crenshaw, *Naval Shiphandling*, 15.
10. Barber, *Naval Shiphandler's Guide*, 181.
11. It should be noted that even when the waterjets are perfectly centered, their off-center location with respect to the keel will produce a twisting moment. Chapter Three will explain how to discern longitudinal and lateral thrust, as well as how the shiphandler can arrange these variable forces to yield a resultant thrust vector that propels the ship in the desired direction.
12. Barber, *Naval Shiphandler's Guide*, 39.
13. Ibid., 38.
14. "Littoral Combat Ship Class: LCS"; "DDG-51 *Arleigh Burke-Class*," *Federation of American Scientists*, accessed 2 September 2014, http://www.fas.org/man/dod-101/sys/ship/ddg-51.htm.
15. Barber, *Naval Shiphandler's Guide*, 147–49.
16. Ibid., 147.
17. Ibid., 145.
18. Ibid., 105.
19. The anchor is at "short stay" when enough chain has been paid out to exceed "up and down" but with the flukes still clear of the bottom. Crenshaw, *Naval Shiphandling*, 99.
20. Barber, *Naval Shiphandler's Guide*, 106.
21. Ibid., 106–107.

CHAPTER 3. WATERJET VECTOR MANAGEMENT

1. Barber refers to this point as the "center of lateral resistance." Barber, *Naval Shiphandler's Guide*, 40.

2. Ibid., 40–43.

3. Since during actual shiphandling force vectors are not specified in knots, the math presented in this chapter is for illustrative purposes only. These solutions are rounded to the nearest tenth of a knot. For those interested in the math, the force vectors for each waterjet form a triangle, with one angle equal to waterjet angle and the second angle equal to 90 degrees, where the longitudinal and lateral vectors intersect. Of course, the third angle is equal to the difference between 180 degrees and the sum of the first two angles. To solve for the individual force vectors: the hypotenuse is the speed equivalent to the force discharged from the waterjet; the longitudinal force vector is equal to the cosine of the waterjet angle multiplied by the hypotenuse; and the lateral force vector is equal to the sine of the waterjet angle multiplied by the hypotenuse.

4. See note 3.

5. This approach to handling waterjets—first introduced by Capt. Robert Butt, USN (Ret.)—gained the favor of the first generation of LCS commanding officers, who appreciated the fine control it afforded and ability to halt momentum toward danger quickly. Captain Butt originally presented this as a "balanced" approach, in that both waterjet force vectors were balanced to port and starboard.

6. For the remaining discussion in this chapter, the mathematics behind waterjet maneuvers is omitted, so that the shiphandler can focus on conquering the concepts of waterjet shiphandling.

7. Some might expect the movement to be equal, but it is proportional when the pivot point is offset from its natural position. For example, if the pivot point has shifted forward because of the application of lateral force aft when the waterjets are put over, the stern will move a greater distance than the bow.

8. Again, Barber refers to this point as the "center of lateral resistance." Barber, *Naval Shiphandler's Guide*, 40.

9. Ibid., 40–43.

10. In each of the following maneuvers, it is possible for the *Freedom* variant to effect the same pivot point movement, but only with the assistance of a tug.

CHAPTER 4. STANDARD COMMANDS

1. Barber, *Naval Shiphandler's Guide*, 53–60; James Stavridis, *Watch Officer's Guide*, 15th ed. (Annapolis, Md.: Naval Institute Press, 2000).

2. Hydrofoils are any thin, plate-like objects designed to produce lift when adjusted relative to the flow of water. The term often refers to vessels that lift out of the water at high speeds, since these vessels are characterized by the controlling surfaces that enable this action. Crenshaw, *Naval Shiphandling*, 16.

CHAPTER 5. PIERWORK

1. Transitioning from a lateral walk to a diagonal crab is not necessarily bad if the ship has plenty of room ahead and astern; ships are often forced to walk diagonally because of insufficient room in the basin to walk laterally into place.

2. As noted in Chapter Three (endnote 5), this method was originally described as a "balanced" approach by Capt. Robert Butt, USN (Ret.).

3. Crenshaw, *Naval Shiphandling*, 381.

4. The "by-half" technique employed here for pierwork was inspired by Barber's description of "bracket and halving" for alongside refueling. See Barber, *Naval Shiphandler's Guide*, 181.

5. The range of options with tugs is somewhat limited with today's powerful Z-drive tugs, since their thrust-to-weight ratio relative to this ship class constrains the power that can be safely used during pierwork. This topic will be discussed in more detail later in this chapter.

6. For the novice shiphandler: the ship should never touch the actual pier structure but rather ride on the fenders that serve as buffers between the hull and the concrete pilings of the pier itself.

7. Crenshaw, *Naval Shiphandling*, 381.

8. The *Freedom* variant has a sail area roughly equivalent forward and aft of the center point, but the principle remains the same. Varying winds can have varying effects on the ship, so the shiphandler must remain aware of the possibility of unintended pivot when alongside a pier hazard. Thanks to Dale Heinken for this insight.

9. Conversely, an offsetting wind may be so powerful that the ship must be kept from being blown across the slip into the opposing pier.

10. While this discussion could, as an academic exercise, describe how to move the pivot point to the stern when docking the ship, the impracticality of that maneuver precludes its inclusion. In this example, it is highly unlikely that the shiphandler would need to back into the slip and then push the bow toward the quay wall; if it were necessary, however, the above principles would apply.

11. The principles for applying force on the bow in directions other than port and starboard are also applicable for tug management, but getting the tug to pull or push at an angle other than perpendicular requires reliable communications between the ship and tug. In any case, it would be nearly impossible for the tug to achieve the same degree of radial accuracy as the bow thruster.

12. This difference is the primary reason that diesel engines are preferable for pierwork, as their throttles can be adjusted fairly liberally without concern that the ship's momentum might spring out of control.

13. This technique should sound familiar to shiphandlers with experience on the *Oliver Hazard Perry*–class frigate, whose thruster and single-shaft/rudder combination can be paired to move the ship laterally.

CHAPTER 6. CHANNEL DRIVING

1. To put distance in the perspective of time, remember that the three-minute rule says that a ship will travel one hundred

yards in three minutes for every knot of speed. A ship driving ten knots, for example, will travel a thousand yards in three minutes.

2. Before using track control, learn the specific settings for your ship. For example, system defaults are set to within fifteen yards of the track and five degrees of the heading, but these settings may be manually changed to suit the commanding officer's preferences.

3. As is the case of reliance on any automated system, shiphandling skills can atrophy over time if track control is used exclusively. The prudent captain will ensure that the team conducts a certain amount of channel driving under manual control to maintain an ability to navigate safely in restricted waters without track control.

4. Barber, *Naval Shiphandler's Guide*, 106.

5. Ibid., 48.

6. Crenshaw, *Naval Shiphandling*, 15.

7. The ship-class squat characteristics data are presented in table format as a function of speed in knots, water depth in feet, and waterway width in yards, resulting in predicted squat depth in feet.

CHAPTER 7. SPECIAL SHIPHANDLING EVOLUTIONS

1. Throughout this chapter, the ship being refueled is referred to as the *ship* and the replenishment ship as the *oiler.*

2. Refer back to Chapter Two, on controllable and uncontrollable forces, for a discussion in more depth on Bernoulli's principle and Venturi forces.

3. Barber, *Naval Shiphandler's Guide*, 181–82.

4. Ibid., 182–83.

5. Ibid., 183.

6. Thanks to Dale Heinken for pointing out this alternate approach. It is possible to conduct the refueling at greater than thirteen knots, but the shiphandler should be aware that increasing speed exponentially increases Venturi forces.

7. Barber, *Naval Shiphandler's Guide*, 179.

8. The by-half technique applied throughout this book was based on Barber's "bracket and halving" system for alongside refueling. He contended that the shiphandler should use bold movements to correct longitudinal positioning at first and achieve finer control by splitting the difference. Ibid., 181.

9. Crenshaw, *Naval Shiphandling*, 158.

10. This last option—using all engines ahead and then all engines back—is possible on the *Freedom* variant, but speeds as high as thirty knots are not advisable. The *Freedom* variant begins planing between twenty-four and thirty knots, so applying enough power to reach thirty knots will put the ship in a planing configuration. It is not recommended that the ship transition out of planing alongside the oiler. The braking method is certainly achievable and safe at speeds lower than twenty-four knots.

11. Barber, *Naval Shiphandler's Guide*, 175.

12. Ibid., 183.

13. To be clear: the by-half method used here is relative to the baseline throttle setting. Increasing the throttle setting to T8 would produce a four-setting differential between T8 and T4, dividing by half would mean a two-setting change to T6. Subsequently, T6 is a two-setting differential above the baseline T4, which divided by half would lead to a one-setting change to T5.

14. Crenshaw, *Naval Shiphandling*, 381.

15. It should be reiterated that the anchor system of the *Freedom* and *Independence* variants are considerably different. In particular, the *Independence* variant only has one shot of chain, which is connected to a spool of cable. The weight of the cable is about two-thirds that of an equivalent length of chain, a weight reduction that contributes to the ship's considerable thrust-to-weight ratio, which in turns permits high speed for this vessel. In contrast, the *Freedom* variant has a traditional anchoring system that is entirely chain. For the simplicity of

this discussion, all references to length of "chain" will apply to both classes, whether the ship is deploying chain or cable.

16. This maneuver was adapted for waterjets from a maneuver originally proposed by Crenshaw for propeller-driven ships. Crenshaw, *Naval Shiphandling*, 101–102.

17. Ibid., 116–17.

18. While a shipboard recovery is always a consideration, a small-boat recovery is generally preferable. Using the small boat allows a rescue swimmer and medical support to evaluate the sailor's injuries, provides the quickest means to get the sailor out of the water, and in the likely event of exhaustion from survival swimming, requires the least effort from the sailor.

19. To be sure, the man overboard will move with the prevailing ocean currents, but from a shiphandling perspective, this movement is minimal.

20. Barber, *Naval Shiphandler's Guide*, 97.

CHAPTER 8. OPEN-OCEAN SHIPHANDLING

1. Ships also experience surge (longitudinal level movement) and sway (lateral level movement), but from the conning officer's perspective in open-ocean shiphandling, these movements are of little significance.

2. John V. Noel, *Knight's Modern Seamanship*, 17th ed. (New York: Van Nostrand Reinhold, 1984), 256.

3. Ibid.

4. This box measures five miles on either side of the track and sixty miles (fifteen miles per hour across four hours) both ahead of and behind PIM.

5. For accuracy, *tacking* is a sailing maneuver for heading generally in the direction from which the wind is blowing; the sailboat zig-zags across the wind, keeping the wind as near to dead ahead as possible, on alternate sides, without being "taken aback" and losing way. This analogy loosely fits here, although here the ship is swinging back and forth across the intended track and keeping the seas off the beam. In both cases, the

vessel is maneuvering down the intended track while keeping the uncontrollable force away from an undesired relative angle.

6. The term *leg* is defined as the distance traveled between the track and the outermost point allowed by the standing orders.

7. In an effort not to distract from the principles of this maneuver in the main text, the mathematic equations are explained here. Anyone trying to reproduce this data, either for validation or to build their own tables, may find it useful to see the methodology explained. It should be noted that distances are solved to the nearest tenth of a nautical mile and that both speeds and time are rounded to the nearest whole number in knots and minutes. In the example, recall that the plotted course is at 270 degrees true with a fifteen-knot speed of advance. Altering course to starboard to 315 degrees true would create a forty-five-degree angle (A) off the plotted course. As the ship is driving away from the track, it is moving down a line that forms the hypotenuse (c) to a right triangle that is also composed of the line (a) perpendicular to the track and along the maximum allowable distance from the track (five miles) and a line (b) along the plotted track. To solve for the distance the ship travels before reaching the lateral limit from the track, hypotenuse (c) is equal to the maximum lateral distance marked by line (a)—5 miles—divided by the sine of angle (A): 5 miles / sin(45) = 7.1 miles. At the point when the conning officer must tack to port, the ship will have traveled a distance parallel to the track marked by line (b), which can be solved by multiplying the hypotenuse (c) by the cosine of angle (A): 7.1 miles × cosine(45) = 5 miles. In determining the speed required to maintain PIM, line (b) becomes the critical factor. If the ship had remained on the plotted track, it would have traveled the distance of line (b)—5 miles—in a time equal to that distance divided by the ship's speed: 5 miles / 15 knots = 20 minutes. The ship must achieve this same distance down the plotted track in the same amount of time had it remained on the plotted track, even though it is steering 315 degrees

true and traveling the distance marked by line (c)—that is, 7.1 miles. So, the speed required to maintain the speed of advance is the actual distance traveled along line (c) divided by the time allotted: 7.1 miles / 20 minutes = 21 knots. Applying just these final equations to another course, where the ship alters course to starboard to 325 degrees true, angle (A) is 55 degrees, line (a) is still 5 miles, hypotenuse (c) is 6.1 miles, and line (b) is 3.5 miles. The time the ship would have traveled 3.5 miles along the track at fifteen knots is fourteen minutes, so the speed required to maintain PIM is: 6.1 miles / 14 minutes = 26 knots.

8. In this example, "by half" refers to the difference between the ordered twenty knots after the maneuver and the subsequent increase to twenty-four knots. If speed had been increased to thirty knots, the by-half sequence would have followed: thirty knots, twenty-five knots, 22.5 knots, and finally twenty-one knots.

9. To be sure, the responsibility for safely navigating the ship *always* rests with its own shiphandling team, but as long as the water ahead is safe, the ship is expected to follow the movements of the guide.

CHAPTER 9. SIMULATOR TRAINING

1. As this is the concluding chapter, the following sections will have the most meaning only after reading the preceding chapters; certainly, Chapter Five, on pierwork, and Chapter Seven, on special evolutions, are prerequisites and should be examined before continuing. Subsequent sections in this chapter will not reexplain concepts such as the toe-in/toe-out method or dynamic thruster management. Although possible to grasp these maneuvers from the text below, the reasoning behind their application will not be clear if the earlier chapters have not been read.

Barber, James Alden. *Naval Shiphandler's Guide.* Annapolis, Md.: Naval Institute Press, 2005.

Crenshaw, R. S. *Naval Shiphandling.* 4th ed. Annapolis, Md.: Naval Institute Press, 1975.

Noel, John V. *Knight's Modern Seamanship.* 17th ed. New York: Van Nostrand Reinhold, 1984.

Stavridis, James. *Watch Officer's Guide.* 14th ed. Annapolis, Md.: Naval Institute Press, 2000.

U.S. Government. *The American Practical Navigator.* Bethesda, Md.: National Imagery and Mapping Agency, 1995.

About the Author

A career naval officer, **JOSEPH A. GAGLIANO** commanded USS *Independence* (LCS 2) and has served on several *Arleigh Burke*–class destroyers. He holds a PhD and master's degree from the Fletcher School of Law and Diplomacy at Tufts University in Medford, Massachusetts; a master's from the U.S. Naval War College in Newport, Rhode Island; and a bachelor's from the U.S. Naval Academy in Annapolis, Maryland.

The Naval Institute Press is the book-publishing arm of the U.S. Naval Institute, a private, nonprofit, membership society for sea service professionals and others who share an interest in naval and maritime affairs. Established in 1873 at the U.S. Naval Academy in Annapolis, Maryland, where its offices remain today, the Naval Institute has members worldwide.

Members of the Naval Institute support the education programs of the society and receive the influential monthly magazine *Proceedings* or the colorful bimonthly magazine *Naval History* and discounts on fine nautical prints and on ship and aircraft photos. They also have access to the transcripts of the Institute's Oral History Program and get discounted admission to any of the Institute-sponsored seminars offered around the country.

The Naval Institute's book-publishing program, begun in 1898 with basic guides to naval practices, has broadened its scope to include books of more general interest. Now the Naval Institute Press publishes about seventy titles each year, ranging from how-to books on boating and navigation to battle histories, biographies, ship and aircraft guides, and novels. Institute members receive significant discounts on the Press's more than eight hundred books in print.

Full-time students are eligible for special half-price membership rates. Life memberships are also available.

For a free catalog describing Naval Institute Press books currently available, and for further information about joining the U.S. Naval Institute, please write to:

Member Services
U.S. Naval Institute
291 Wood Road
Annapolis, MD 21402-5034
Telephone: (800) 233-8764
Fax: (410) 571-1703
Web address: www.usni.org